CAMBRIDGE COUNTY GEOGRAPHIES
SCOTLAND
General Editor: W. Murison, M.A.

PEEBLES
AND
SELKIRK

Cambridge County Geographies

PEEBLES

AND

SELKIRK

by

GEORGE C. PRINGLE, M.A.

Rector, Burgh and County High School, Peebles

With Maps, Diagrams and Illustrations

Cambridge:

at the University Press

1914

CAMBRIDGE UNIVERSITY PRESS
Cambridge, New York, Melbourne, Madrid, Cape Town,
Singapore, São Paulo, Delhi, Mexico City

Cambridge University Press
The Edinburgh Building, Cambridge CB2 8RU, UK

Published in the United States of America by Cambridge University Press, New York

www.cambridge.org
Information on this title: www.cambridge.org/9781107651845

First published 1914
First paperback edition 2013

A catalogue record for this publication is available from the British Library

ISBN 978-1-107-65184-5 Paperback

CONTENTS

good—" They inquired almost timidly, remembering that the man in front of them had become prosperous through the workings of this phase of his philosophy.

"Mark my words."

He rose slowly and carefully put down the plate on which his tea had been served. "Well, I've got to get back home. Somebody'll be dropping by for some gasoline pretty soon." He hesitated for a moment, then turned to Mrs. Frake.

"Well, Mrs. Frake, I reckon you're mighty glad you went to the Fair, aren't you?"

"Of course I am. What makes you ask that?"

"I guess you two youngsters are mighty glad, too?" He looked at them, and the ghost of a smile touched the corners of his mouth. Then, suddenly, between Wayne and the Storekeeper and Margy and the Storekeeper there passed a secret glance of understanding, and a smile that was a promise.

They laughed. "We certainly are."

He shrugged his shoulders and turned away. "Abel, there's a little bit of business I'd like to talk with you if you could come out to the car with me."

Abel grinned. "Be right back, folks." Together, the two men went out to the Storekeeper's car.

"You're getting fat, Abel, just like Blue Boy." He felt in the bottom of the car and drew out the crank.

Then he reached into his vest pocket and handed Abel a five-dollar gold-piece. "Well, here you are. I just hope there's nothing funny about it all."

"Thanks," said Abel, gravely.

The Storekeeper tried the crank. After the third or fourth attempt the motor wheezed twice. "Why in the world," said Abel, "don't you get a car with a self-starter? You've got enough money in the bank to buy fifty cars."

"More things to get wrong," said the Storekeeper, panting, and putting his shoulders into another twist. The car wheezed twice, again, paused, wheezed, coughed reluctantly, and suddenly began roaring as the Storekeeper threw down his throttle.

The Storekeeper, protruding from the rackety little machine like a Buddhist idol from its shrine, turned painfully and carefully around the wide carriage-yard and set off down the road.

Once on the river road, the Storekeeper shook his head and laughed quietly to himself. The car was running at a dreadful speed of perhaps twenty miles an hour, so he throttled it down to fifteen. Shadows were gathering in the willows; the birds were already sleepy and the cicadas were dismally awake. The Storekeeper contemplated the road before him thoughtfully.

Twice he shook his head. Once he nodded and chuckled. Then his brows drew up in serious thought and he looked at the blue, still faintly lighted heaven, which shone between the tree-tops. "I wonder," he said to himself.

At the bottom of the Dooley Hill he reached meditatively into his coat pocket and drew out a cigar, still considering the clever trick by which They had cheated him of ten dollars. He chewed the cigar solemnly until he had climbed to the top of the hill and then he reached into another pocket for a match. There was no match. Slowly, carefully, he searched through all his pockets.

At last, in a side-pocket of the car, he found one match, a fat, dry, well-made match, with a pink head, white-tipped. He was about to strike it when a thought seized him. Painstakingly, he repeated his search. There was no doubt about it, in his fingers he held the only match in the car.

He threw it out of the car. "It would go out," he murmured.

A mile farther on he glanced again at the lighted blue above the trees and sardonically, but not unpleasantly, he chuckled.

ILLUSTRATIONS

MAPS

The illustrations on pp. 2, 12, 93 and 107 are from photographs by Messrs J. Valentine and Sons; those on pp. 4, 8, 19, 21 (St Mary's Loch), 40, 56, 73, 78, 94, 109, 112, 116, 127, 130, 135, and 144 are from photographs (a number of which were specially taken for this book) by Mr A. R. Edwards, Selkirk; those on pp. 21 (Entrance to St Mary's Loch) and 70 are from photographs by Mr Colledge, Innerleithen; those on pp. 29 and 31 from photographs, taken by Mr Colledge, of fossils lent by Mr George Storie, a former pupil of the author; those on pp. 37 and 59 from photographs by Mr J. Ward, Peebles.

Thanks are due to the Tweeddale Society, through Mr J. Walter Buchan, for the use of blocks from which the illustrations on pp. 16, 97, 103 and 108 are reproduced; to Dr John Bartholomew, Geographical Institute, Edinburgh, for permission to reproduce the illustration (adapted) on p. 26 (fig. 1); to Messrs Ballantyne and Co., Peebles, for permission to reproduce those on p. 63; to the Directors of the South of Scotland Technical College, Galashiels, through Dr Oliver, for the use of the block for the illustration on p. 64; to Messrs R. Smail and Sons, Innerleithen, for permission to reproduce those on pp. 68 and 106; to Mr T. Craig Brown, Selkirk, for permission to reproduce the map on p. 81 and the illustration on p. 117; to the Society of Antiquaries of Scotland to reproduce the plan on p. 83; to

Dr C. B. Gunn, Peebles, for the use of the blocks for those on pp. 84, 86, 89 and 121; to Mr Ross for permission to reproduce from McGibbon and Ross' *Castellated Architecture of Scotland* the plan on p. 104; to Mrs Andrew Lang for kindly supplying the portrait on p. 133, and to Mr Allan Smyth of the *Peeblesshire Advertiser* for the use of the block for the illustration on p. 139. The maps on pp. 99 and 114 and the sketches on p. 100 were made by J. Connel Pringle, who also adapted that on p. 26 (fig. 1).

For useful information and suggestions the author desires to express his obligations to: Lord Glenconner of Glen; Mr T. Craig Brown, Selkirk; Mr James Sanderson, Woodlands, Galashiels; Dr Oliver, Galashiels; Mr J. Ramsay, Board of Agriculture for Scotland; Mr Watt, Scottish Meteorological Society; Messrs Leslie and Reid, of Edinburgh and District Water Trust; the Rev. Wm. McConnachie, Lauder; Mr G. Constable, Traquair; Mr J. Ramsay Smith, Peebles; the late Mr R. S. Anderson, Peebles; Mr Bartie, Selkirk; Dr C. B. Gunn, Peebles; Mr Herbertson, Galashiels; Mr M. Ritchie, High School, Peebles; Messrs W. and T. Paterson, Crookston, Peebles; Mr Geo. Wilkie, Mr Wm. Sanderson, Mr W. Johnstone, Peebles, and others.

NOTE

In other volumes of the series dealing with two counties, e.g. *Argyllshire and Buteshire*, each county is treated separately. This method, however, was found less suitable in the case of Peebles and Selkirk, which have been treated for the most part as a single area.

1. County and Shire. The Origin of Peebles and Selkirk.

The word *shire* is of Old English origin and meant office, charge, administration. The Norman Conquest introduced the word *county*—through French from the Latin *comitatus*, which in mediaeval documents designates the shire. *County* is the district ruled by a count, the king's *comes*, the equivalent of the older English term *earl*. This system of local administration entered Scotland as part of the Anglo-Norman influence that strongly affected our country after the year 1100. Our shires differ in origin, and arise from a combination of causes—geographical, political and ecclesiastical.

The first known sheriff of Selkirk was Andrew de Synton appointed by William the Lyon (1165–1214); and there were sheriffs of Peebles in the same reign. In 1286 Peebles had two sheriffs, one holding his courts at Traquair, the other at Peebles—the two courts being amalgamated about the year 1304. In Alexander II's reign Gilbert Fraser was sheriff of Traquair, while in the reign of Alexander III Sir Simon Fraser was sheriff of Peebles and keeper of the forests of Selkirk and Traquair.

But these counties were more familiarly known by other names. In State Documents Peebles was frequently called Tweeddale (Tuedal), and Selkirk, Ettrick Forest or the Forest. Even in Blaeu's Atlas (1654) the inscription on the map of the two counties is: "Twee-Dail with the Sherifdome of Ettrick Forest, called also Selkirk."

Peebles from the West

Ettrick Forest—sometimes, and presumably later, Selkirk Forest—was, however, much more extensive than the present Selkirkshire.

The name *Peebles*, older form *Peblis*, is generally regarded as derived from the British word *pebyll*, tents, place of tents. *Selkirk*, old spelling *Scheleschirche*, is taken to mean the kirk of the shieling.

No doubt the counties came into existence as convenient districts determined mainly by natural conditions as rivers, mountains, forests, for the administration of local and national affairs. Peebles corresponded to the Vale of the Tweed from the source of the river till it approaches the region of its first large tributary, the Ettrick from the Forest, the watershed between the Tweed and the Ettrick forming a natural boundary. The Shire of the Forest was a distinctive area at first marked out and set aside as a hunting preserve for the Scottish kings. As political and social conditions have changed, these counties have also changed in shape and to some extent in size.

2. General Characteristics.

Peebles and Selkirk are entirely inland counties; but they are not so cut off from the sea as not to be affected by the outer world and as not to affect it. No region on the face of the earth, not even Greece excepted, has been more "besung" than the Border Ballad district embraced in Selkirkshire. Burns says "Yarrow and Tweed to monie a tune owre Scotland rings" and the poetry of the district is without doubt its chief claim to distinction. The Tweed or woollen industry has rendered these counties no less famous in the sphere of commerce.

It is not necessary to assume that spiritual and mental characteristics are entirely due to material causes. If the people of the Forest and of the Uplands of Peebles and

Selkirk were brave and romantic it does not follow that it was the Forest and the Uplands that made them so. It was probably an initial endowment of the spirit of adventure and love of freedom that drove many of the early inhabitants into these fastnesses where even the king as well as foreign foes hesitated to intrude. But the natural conditions of the Forest had undoubtedly a great influence on the thoughts, emotions and occupations of its inhabi-

Selkirk from the North-West

tants—the conditions: (1) that the counties belong to the Southern Uplands, a district noted for its suitability as a pastoral region and for its picturesque beauty; (2) that they are included in the district of the middle marches over which the tide of war ebbed and flowed for centuries.

It was natural that a region in which King James IV at one time had as many as 10,000 sheep and from which much wool was exported to Flanders should have woollen factories as at Galashiels, Selkirk and Peebles. But besides the sheep there were cattle in the meadows, and beasts in the Forest, whence oak bark was obtained for tanning. So that there was also leather in abundance and up to the end of the eighteenth century Selkirk was more famous for its shoe-making than Galashiels for its woollen manufacture.

Although the counties took more than their share in the extension and improvement of agriculture in the eighteenth century, yet owing to the hilly nature of the region and the consequent thinness of the soil, the counties, except in the north-west of Peeblesshire, have remained chiefly pastoral. The present outstanding features of the district therefore are sheep-farming and woollen manufactures. But at the time when planting became fashionable in Scotland, in no part of the country did so much planting of timber take place, as in the counties of Selkirk and Peebles. Indeed, previous to the extension of railway lines into the counties it was considered that this planting had been overdone. In the vicinity of the county towns and in such districts as Bowhill, in Selkirkshire, and Cademuir Hill, in Peeblesshire, a great change has been effected in the appearance of the landscape by the planting of woods and forests, mainly pine. At the time referred to numerous estates particularly in Peeblesshire were purchased by wealthy merchants and professional men and vast sums of money expended on laying out policies, on building, draining and planting. One estate in particular, the property of the Earl of Islay, afterwards the third Duke of Argyll, obtained its name, "The Whim," in token of the excessive outlay in converting a wild morass into a pleasure ground. From its romantic associations, picturesque attractions, and its proximity to Glasgow and Edinburgh, wealthy proprietors have helped to make Peeblesshire the county with the highest valuation (12·5) per head of the population in Scotland. Selkirkshire, however, has remained chiefly in the hands of one or two of the great nobles—the Buccleuchs and the Napiers; and consequently the ratio of its valuation (6·5) to its population has not increased to the same extent.

3. Size. Shape. Boundaries.

The area of Peeblesshire is 222,240 acres of land and 1048 acres of water. Selkirkshire has 170,793 acres of land and 1796 acres of water, and is therefore about three-quarters the size of Peebles. Peebles could be contained in Inverness more than twelve times, and could itself contain Clackmannan more than six times. It comprehends one eighty-seventh part of the land and water of Scotland.

Peeblesshire is roughly triangular in form. The longest side stretches from Borestone in the north of the parish of Linton to the Great Hill where the Coreburn takes its rise, on the southern boundary between the parishes of Tweedsmuir and Moffat. A line drawn through Great Hill and Dollar Law to Thornilee in a north-easterly direction marks the direction of the south-eastern boundary between the two counties. The third and shortest side of the triangle runs north-west from

Thornilee to Borestone. The Tweed basin with its tributaries fills up this triangular area, the sides of which converge towards its south-eastern apex.

On the west Peebles marches with Lanark, on the north with Midlothian, on the south with Dumfries, and on the south-east with Selkirk.

With the exception of the portion which projects in a south-westerly direction into Dumfriesshire, the outline of the county of Selkirk may be described as an ellipse or oval of irregular outline, with its main axis lying north-east and south-west. The greatest length along the main axis from Capell Fell to Galashiels is twenty-seven miles. The greatest breadth from Dear Heights in the north of the Caddon division of the county to Hangingshaw Hill north of the Ale Water is about the same.

Selkirk marches with Peebles on the north-west, with Dumfries on the south-west, with Roxburgh along the eastern curve, and with Midlothian on the north.

Before 1892, when the Boundary Commission for Scotland was appointed, several detached portions of the one county lay within the other. The parish of Lyne in Peeblesshire had previously been joined with that of Megget in Selkirkshire to form one parish, although separated each from the other by the whole length of Manor Vale and parish, a distance of fully fourteen miles. The Commissioners ordered that Megget should form part of the parish of Yarrow in the county of Selkirk. Similarly the portions of the parishes of Peebles and Innerleithen, which used to be in the county of Selkirk, are now in the county of Peebles. A detached portion of Yarrow parish, about 2166 acres, surrounded by the parishes of Peebles, Innerleithen and Traquair, was united to Traquair parish (which the Yarrow portion had divided into two) in the county of Peebles. The parish of Culter no longer exists. From 1801 to 1851 it was returned as wholly in Lanark; from 1851 to 1891 part of it was returned in Peeblesshire. In 1891 this portion was transferred to the parish of Broughton, Glenholm and Kilbucho.

Yarrow Kirk and Manse

The Commission had also to deal with parishes partly in Selkirk and partly in Roxburgh and Midlothian. Roberton parish in the east, which used to be partly included in Selkirk, is now entirely within the county of Roxburgh. Portions of the parishes of Ashkirk, Selkirk and Galashiels, partly in Selkirk and partly in Roxburgh, were transferred to the county of Selkirk. The large and growing town of Galashiels close to the borders of Roxburgh and Selkirk had to extend its boundaries eastwards; and the Commissioners decreed that the portion of Melrose parish in the county of Selkirk should become part of the parish of Galashiels and of the county of Selkirk. Still later, in 1908, another portion of Melrose parish was annexed to the burgh of Galashiels for drainage purposes, and in 1911 annexed to the parish of Galashiels.

The anomalies were not, however, all removed. The parish of Stow is situated partly in the county of Edinburgh and partly in the county of Selkirk. The Selkirkshire portion, known as Caddonfoot, is of large area with a population almost wholly agricultural; and as there were reasons against bringing Edinburgh down to the Tweed, as well as against making Caddonfoot part of Galashiels, this portion of Selkirkshire was kept within the parish of Stow. In 1898, however, by order of the Secretary for Scotland, it was formed into the parish of Caddonfoot in the county of Selkirk together with portions of the parishes of Selkirk, Galashiels and Yarrow.

These changes do not affect the ecclesiastical parishes.

4. Surface and General Features.

The part of southern Scotland known geographically as the Southern Uplands, a region now cut and carved into valleys and watersheds, was formerly a lofty tableland. A line drawn through Penicuik, Galashiels and Melrose, where the Tweed leaves the Uplands and enters the plain, and another line from the Moffat hills to Melrose along the ridge separating the Ettrick from the Teviot, will practically cut off that portion of the Uplands which contains the counties of Peebles and Selkirk. The whole of this portion is filled with hills the tops of which are flattened or rounded, the sides smooth, and (except in the highest parts, where peat and heath are frequently found) covered with grass, crags and rocks being rare. This region is in the main pastoral and has hardly any cultivated ground except along the haughs or on the lower slopes of the hills. The most extensive areas of hill peat are found on the Moorfoots on the high ground overlooking the Leithen water and also on the Manor hills to the south-west. These uplands are bare of any natural wood, but in the lower reaches of the Tweed and its longer tributaries, many of the hills are clothed to their summits with woods and plantations, most of them planted within the last hundred and fifty years. The district south of the Tweed including all Selkirk and more, was at one time the Forest of Ettrick.

Starting from the central mass of the Uplands in which rise the Tweed, the Annan and the Clyde, the trend of the valleys, and, therefore, of the ridges between them, is towards the north-east, till we come to the bank of the Tweed, when we are met with ridges on the north side with a trend to the south-east. The former valleys are called longitudinal, because in a line with the strike of the strata, and the other transverse, because at right angles to the strike. Examples of longitudinal valleys are: Ale, Ettrick, Yarrow, Holms, Tweed (to Broughton),

Manor, Quair; of transverse valleys: Biggar, Lyne, Eddleston, Leithen, Walkerburn and Gala. These ridges and rounded masses approach so near and interfold and overlap on each bank so closely that, apart from other proofs, it is apparent that the whole region has at one time been a plateau which the Tweed and its tributaries with other agencies have scoured and grooved and rubbed down into what resembles a rounded, billowy ocean.

The only comparatively level part within the two counties is the district towards the north, stretching between the Moorfoots and the Pentlands from a low watershed, sloping away on the one side towards the shores of the Firth, and, on the other, towards the south-west into the Clyde valley. A flattish range of hills between Eddleston and Lyne waters divides this vale in two, the western portion running north-east and south-west between the Pentlands and the north-western edge of the Southern Uplands. This plain varies in breadth from four miles at Auchencorth in Midlothian to less than one hundred yards in places between Romanno Bridge and Skirling. The surface is arable, well cultivated and wooded, with stretches of moorland towards the Pentlands.

A line from Leadburn through Romanno, Skirling, and Culter separates these two distinctly different regions, the one lowland and arable, the other upland and pastoral. This line coincides with a great "fault" between two different geological formations.

Six sections may be distinctly marked out in this upland region. The first is Selkirkshire, with its parallel ridges lying north-east and south-west from the high central mass culminating in Capel Fell and Ettrick Pen and forming the watersheds between the Tweed and the Yarrow, Yarrow and Ettrick, Ettrick and Teviot. Each of these valleys has its south-western end wild, mountainous and treeless; its middle region pastoral, with grassy or heathery rounded hills and occasional clumps of dark pines

Galashiels

near the farm houses; its lower end a region of wood and hill, pasture and arable land. The second section is bounded by the ridge between Peebles and Selkirk on the south and the Tweed on the west and north from its source to Galashiels. This area is occupied by the parallel masses separating Tweed and Manor, Manor and Quair, and other lesser streams till Ettrick meets Tweed. Here, as before, the valleys have the three-fold character of wilderness; pastoral; mixed pastoral, woodland and arable. Thirdly, there is the triangle bounded by the Eddleston Water, the Tweed, and the boundary line through the Moorfoots—a high region, several summits being over 2000 feet. Intersected by the transverse valleys of the Leithen and the Walkerburn, it consists mainly of pasture and moorland. In the extreme north above Portmore Loch the ground is low and forms part of the valley between the Moorfoots and the Pentlands. Round Portmore the ground in the lower reaches near Eddleston is well wooded. The fourth section, mainly pastoral, is an undulating region, the chief heights being the Meldons between the Eddleston Water and the Lyne. The fifth division consists of the heights behind Stobo and Broughton, bounded on the north by the Tarth and on the west by the Broughton Burn. Beyond that again is the last section, the agricultural region stretching from Skirling, Romanno and Leadburn to West Linton and merging into the moorland towards the Pentlands on the north-west.

5. Watershed. Rivers. Lochs.

The Southern Uplands is a land of waters and watersheds. "A hill, a road, a river" was an English traveller's terse description in the eighteenth century. Although now in many parts woods and forests cover the slopes of the hills and fringe its roads and rivers, hills and rivers still remain its prominent features.

The region was at one time an undulating plateau from whose higher parts streams flowed in all directions. The Tweed, therefore, in a real physical sense has made these hills; and not only made them, but also established them; for, by the Tweed along with other sub-aerial influences they have been made into hills of "stable equilibrium." Without the river, then, the region would be meaningless not only to those who take delight in its beauty and in its historical associations, but also to those who study its physical configuration.

The general slope of the plateau is towards the south-east. Hence the Tweed in its course from Peebles to Berwick, with its tributaries the Lyne, the Eddleston, the Leithen and the Gala, flows to the south-east. As the course of these rivers was originally determined by the slope of the ground they are called *consequent*, from which we infer that they are the oldest rivers of the country. This agrees with the fact that in former geological days, a great river crossed the country from the region of Loch Fyne to the North Sea, by the present Clyde Valley, and by the present Tweed Valley, which it entered near Biggar. Various changes occurred, which ultimately resulted in the Clyde and the Tweed as we know them. On the other hand, the Tweed from Tweedsmuir to Drummelzier, and the tributaries, the Holms Water, the Yarrow, the Ettrick, all flowing north-east, must have been formed *subsequently* to the time when the course of the main rivers was settled—probably after the great ice age. Hence such rivers are called *subsequent*.

The Tweed (103 m.) rises at Tweedswell, 1250 feet

above sea-level. After a north-easterly direction as far as Peebles it turns east-by-south through Peebles and Selkirk till it meets the Ettrick; then turning north, it receives the Gala and a little below Galafoot enters the county of Roxburgh. Its total course through Peeblesshire, from Tweedswell to Scrogbank, is 40 miles, and through Selkirkshire from Scrogbank to the railway bridge, between Galashiels and Melrose, 10 miles. In Tweedsmuir the only tributary of any size is the Holms Water, which unites with the Biggar Water and the Broughton Burn. The hills in the south-west of Peeblesshire have their highest summits lying to the east and north of Tweedswell, on the boundary line between Peebles, Dumfries and Selkirk. These are Hart Fell (2651), Loch Craighead (2625), Broad Law (2723), and Dunlaw (2584). It is in these hills that the Tweed receives such streams as the Fruid, the Talla (the catchment area of Talla Reservoir) and the Stanhope. After a course of 15 miles it enters, below Rachan, the haughlands of Drummelzier, the widest part of the Tweed valley above Melrose. Into this plain the valleys of Biggar and Broughton converge from the west. Near Drummelzier church the Tweed is joined by the Powsail Burn from Merlindale. The rhyme, attributed to Thomas of Ercildoune,

> "When Tweed and Powsail meet at Merlin's grave,
> England and Scotland shall one monarch have,"

is said to have been fulfilled on the day that James VI became James I of England.

Talla Linns

Eastwards, beyond Dawyck and Stobo with their beautiful woods, the Tweed receives the Lyne from the north-west of the county. More than a mile further on it meets the Manor Water, with a course almost parallel to that of the Tweed, the heights between the two streams comprising Dollar Law, Pykestone, and the Scrape. The river has now arrived at the picturesque pass of Neidpath, through which it joyously forces its way above the town of Peebles (see page 2). Here it is joined on the north bank by the Eddleston Water, which flows almost due south from Leadburn heights through a beautiful upland valley. Haystoun valley to the east and south of Peebles, through which flows Haystoun Burn, shows evidence of having once formed the old bed of the river, which flowed from a large lake stretching beyond Neidpath and Cademuir, well up towards Drummelzier. Once the water at Neidpath had worn down the shaly rock sufficiently to drain the lake, the course in the Haystoun valley gradually shrank from one lake with a river current through it to a series of small lakes joined by a narrow stream. These lakes existed up to 1823, when they were drained and the cutting exposed the bottom of the old lake.

At Peebles the river has fallen 800 feet. Between Peebles and Innerleithen on the north and Traquair on the south bank, the river winds through a beautiful valley diversified with gently sloping and interfolding hills, natural forest, wooded parks, green haughs with glimpses of cattle cooling their limbs at summer noon in shaded pools, of ancient peel towers perched on rocky slopes, or of modern mansions gleaming through the trees. Near

Traquair House the Tweed was diverted northwards for a distance of two miles from its old course. This part of the river used to be known as the "New Water." The Quair, which here joins the Tweed on the south bank, small as it is, is one of the historic streams of Scotland. It runs parallel to Manor and in its romantic valley stand the church of Traquair, and the mansion house of the Glen. Half a mile further on, the Tweed is joined by the Leithen Water flowing down a steep pastoral valley from the Moorfoot Hills.

About one mile west of Elibank Castle the Tweed becomes the boundary between the counties, and half a mile below Thornilee station it enters the parish of Caddonfoot in Selkirkshire. Nearly three miles to the south-east it passes Ashestiel, opposite which the highroad strikes over the hill to Clovenfords. South of Clovenfords the Caddon Water enters the Tweed at Caddonfoot. Neidpath hill on the opposite bank turns the current to the south towards Yair House, where the river rushes over a series of rocky boulders called "Yair Trows." Here Sir Walter Scott used to "leister" salmon. The Tweed is now joined by the Yair Burn. On the left bank a little further on stands Fairnilee, and below Sunderland Hall the Ettrick from Selkirk, the largest tributary, enters on the south bank. Then passing Abbotsford and receiving the Gala from the Moorfoots, half a mile beyond Galafoot, the Tweed enters Roxburgh, where it finally leaves the Southern Uplands for the wide plain between the Cheviots and the Lammermuirs.

The Ettrick (30 miles) rises in Ettrick Pen. Its

valley is larger and wider than Yarrow's, and, in its upper reaches, wilder and more picturesque. Only a few of its numerous tributaries can be noted. On the right is the Tima, from Eskdalemuir; on the left the Kirkburn and the Scabscleuch, with a road over to Yarrow. Further down is the Rankleburn with the Buccleuchs, Easter and Wester, whence the family took their title. On the north

Ettrick Pen

is Tushielaw Tower, home of Adam the Reiver. Three miles on Ettrick receives Gilmanscleuch Burn on the left, and then the Dodhead Burn, scene of Jamie Telfer's "Fair Dodhead," on the right. Northwards through Ettrick Shaws the scenery is picturesque, Ettrick rushing through thick plantations over its rocky bed till Ettrick Bridge End is reached and the old bridge of Wat o'Harden.

On the right is Oakwood Tower, on the left Bowhill, where now Ettrick sweeps with opposing curve to meet Yarrow round the Carterhaugh, scene of "Young Tamlane." Thence northwards Ettrick passes Lindean and enters Tweed.

The Yarrow, rising near Birkhill, flows through the Loch o' the Lowes and St Mary's Loch, into which also flows the Megget. On the shores of the loch are Tibbie Shiel's Inn, the Rodono Hotel and, near the high road, Perys Cockburn's Grave. Further down the valley are St Mary's Chapel, Dryhope Tower, Blackhouse Tower—all three famous in tragic ballad. Still further on, the Gordon Arms, Mount Benger, Yarrow Manse, "the Dowie Dens," are passed, till Hangingshaw with its noble trees, Broadmeadows, once the desire of Walter Scott's heart, Bowhill and Philiphaugh, all beautifully wooded, proudly welcome Yarrow home as it ends its course in Ettrick, east of Carterhaugh.

St Mary's Loch and the Loch o' the Lowes, originally one, stretch along the valley of the Yarrow for about two-thirds of their length. The Oxcleugh Burn and the Whitehope Burn have pushed their deltas out from the shore until they have eventually cut the loch into two parts, and raised the water level of the upper part (the Loch o' the Lowes) so that it drains across the lowest part of the encroaching delta to the lower sheet of water (St Mary's). The Megget is also extending its delta towards the shore below Bowerhope hill, the distance between the two shores being now only a quarter of a mile. In time, therefore, there will be three lochs

Entrance to St Mary's Loch

St Mary's Loch
(*Delta formation at Cappercleuch*)

instead of two. The lochs are remarkably free from vegetation :

> "nor fen nor sedge
> Pollute the pure lake's crystal edge,
> Abrupt and sheer the mountains sink
> At once upon the level brink."

The tableland between Ettrick and Teviot has a chain of small lakes representing evidently an ancient river bed. Some of them contain deposits of shellmarl. These are Kingside Loch, between Selkirk and Roxburgh ; Crooked Loch, a mile further east ; Clearburn Loch ; Hellmuir Loch ; Shaws Lochs (Upper and Under) ; Alemuir Loch. Another row of lochs parallel with these extends for a distance of six or seven miles through Ashkirk, northwards to Selkirk—Shielswood Loch, Essenside Loch, Headshaw Loch, and the Haining Loch. The Haining Loch is an example of a loch tending to disappear through the growth of vegetation. A fresh-water weed, not met with in any other British lake, was discolouring the loch and threatening to fill it up. In 1911 an attempt was made to kill the weed by a solution of sulphate of copper, and so far the experiment has been successful. Cauldshiels Loch, three-eighths of a mile long, one-eighth of a mile wide, 80 feet deep and 780 feet above sea-level, is situated near the boundary line between Selkirk and Roxburgh, with Abbotsford Estate on one of its sides.

The lochs in Peeblesshire are neither so numerous nor so large as those in Selkirkshire. Gameshope Loch, in the very heart of the Peeblesshire wilds, is the highest sheet of water in the south of Scotland, being between 1750 and 2000 feet above sea-level. Talla Reservoir is an artificial barrier loch, forming one of the Edinburgh and District supplies. The surface area of the Reservoir when full is 300 acres; the daily quantity of water available is ten million gallons. Slipperfield Loch, near Broomlee station, 1½ miles in circumference and 845 feet above sea-level, is an example of a lake formed in the upper or stratified drift common in the hills between Linton and Dolphinton, where sand and gravel undulate into hummocky and conical forms and sometimes, as here, enclose pools of water. Portmore Loch, 1000 feet above sea-level, is surrounded by the beautiful woods of Portmore. The North Esk Reservoir, on the boundary between Midlothian and Peebles, and about one mile north of Carlops, supplies Edinburgh and District with water.

6. Geology.

Geology is the science that deals with the solid crust of the earth; in other words, with the rocks. By rocks, however, the geologist means loose sand and soft clay as well as the hardest granite. Rocks are divided into two great classes—igneous and sedimentary. Igneous rocks have resulted from the cooling and solidifying of molten matter, whether rushing forth as lava from a volcano, or, like granite forced into and between other rocks that lie below the surface. Sometimes pre-existing rocks waste away under the influence of natural agents as frost and rain. When the waste is carried by running water and deposited in a lake or a sea in the form of sediment, one kind of sedimentary rock may be formed —often termed aqueous. Other sedimentary rocks are accumulations of blown sand: others are of chemical origin, like stalactites: others, as coal and coral, originate in the decay of vegetable and animal life. For convenience, a third class of rocks has been made. Heat, or pressure, or both combined, may so transform rocks that their original character is completely lost. Such rocks, of which marble is an example, are called metamorphic.

The crust of the earth, in cooling, has contracted into ridges and hollows. The ridges have been worn off and sometimes turned over. Hence it is possible to examine thousands of feet of the earth's crust from its upturned edges. When one system of rock is laid down regularly and continuously upon another the two systems are said to be *conformable*. But if the rocks of the underlying system have been elevated and tilted, or if its surface has been worn away before the younger system has been deposited upon it, the two systems are said to be *unconformable*. From the order of the strata, from their conformity or nonconformity, and from the characteristic fossils belonging to the various divisions and sub-divisions, we learn that the rocks of Peebles- and Selkirkshires belong to the Palaeozoic or Primary group of rocks, that they are younger than the Cambrian, and older than the Old Red Sandstone and than the Coal Measures of the same group.

When a section of the earth's crust sinks down in a gap or fracture so that the beds are displaced on each side of the fracture the displacement is called a *fault*. Two faults run north-east and south-west forming the boundaries of the coalfields of Central Scotland. The Southern Uplands lie to the south of the southern line of fault. That is to say, the Old Red Sandstone and Carboniferous strata of the Midlothian coalfield lie up against the Ordovician of Peeblesshire (Fig. 1, p. 26).

The surface of the Uplands is greatly wrinkled and contorted. The ridges of strata are called *anticlines*, the hollows, *synclines*, see Fig. 2 (p. 26) where *a*, *a*, are anticlines, *b*, *b*, synclines. The anticlines may sometimes be so folded over that younger strata lie below older. This has happened in the case of the Birkhill Shales and has consequently made reading of the geological record a difficult task.

In the Ordovician period the strata were laid down in a sea which covered Wales and southern Scotland. In this sea lived plants, and animals of simple form like graptolites, trilobites, corals and starfish. It was a period of intense volcanic activity, and igneous rocks found their way through rents and fissures. A strip of about seven miles broad in the north of Peeblesshire belongs to the Ordovician (Lower Silurian) period. But the greater part of Peeblesshire and the whole of Selkirkshire belong to the period which followed, namely, the Silurian proper (Upper Silurian). This latter period was characterized by the deposition of sediments and limestones in a shallow, quiet, and wide spreading sea; and the life of the period marks a great advance on that of the one previous. For certain forms of insects and fish,

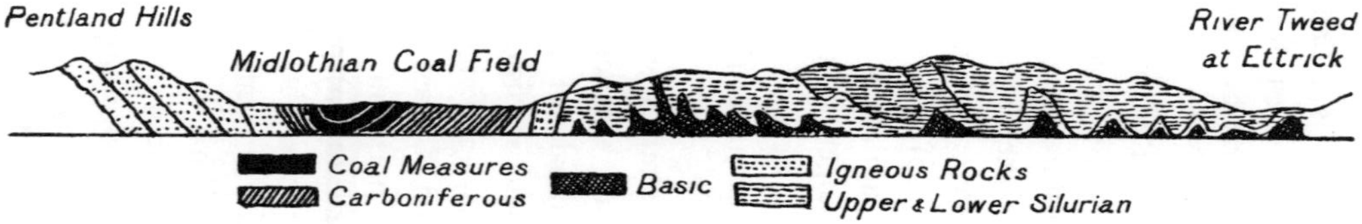

Pentland Hills

Midlothian Coal Field

River Tweed
at Ettrick

Coal Measures
Carboniferous
Basic
Igneous Rocks
Upper & Lower Silurian

Fig. 1. Section from Pentlands S.E. through Midlothian coalfield and Southern Uplands to Tweed at Ettrick

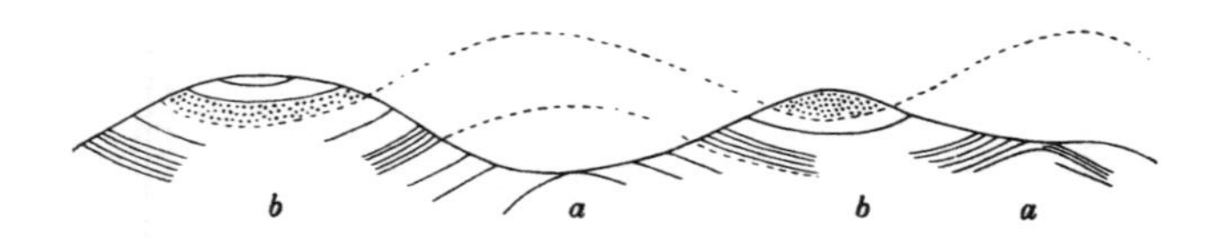

Fig. 2. Hills of Synclinal formation

and the first representatives of the backboned animals, now began to appear. Till 1852 it was thought that the Silurian rocks of the district were destitute of fossils. But James Nicol, son of the minister of Traquair, showed that greywacke (the older name for Silurian rock) was fossiliferous. Later, Lapworth, then a teacher at Galashiels, established a distinction between the two systems (Upper and Lower Silurian); and, because the latter system is best developed in Wales, named it Ordovician after an ancient Welsh tribe of that district.

After these rocks became land, the downthrow in the trough fault of Central Scotland caused a ridging up of the Southern Uplands into a real mountain range from Girvan to Dunbar, so that the rocks of Peeblesshire, the general dip or inclination of which is N.N.W., plunge in that direction beneath the great Carboniferous basin of southern Scotland not again to reappear till they emerge in a much narrower band under the Grampians. In course of time, however, these mountains of elevation were worn down by sub-aerial forces and the process of denudation was assisted by the fact that the strata of these mountains were anticlines, that is to say, sloped away from the axis of elevation (Fig. 3), whereas in mountains built up of synclines, the strata would slope towards the axis of elevation (Fig. 4) and the mountains would therefore be of more stable equilibrium.

The process may be further illustrated by Fig. 5; from which we may see that the masses *A*, *C*, *E* would be gradually worn down to an undulating plain, which, having been once more raised to a high plateau of about

3000 feet, the sub-aerial forces renewed their work with increased vigour till the hills and valleys of the two counties assumed practically their present outlines. In this way the hills of Peebles and Selkirk became hills of circumdenudation, *i.e.* they were, so to speak, dug out not raised up like mountains of elevation. They also became hills of synclinal formation like *B* and *D*, and their valleys valleys of erosion like *a*, *c* and *e*,

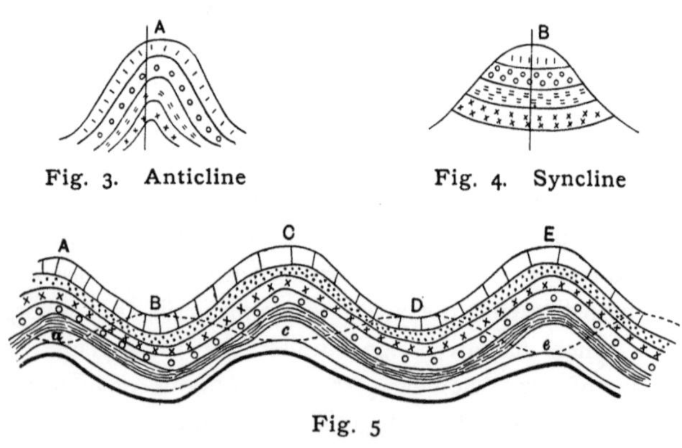

Fig. 3. Anticline

Fig. 4. Syncline

Fig. 5

A B C D E surface before, and a B c D e surface after long period of denudation

where, as the erosion continued, the older rocks would be exposed. Thus "the valleys were exalted and the mountains were laid low."

After the Silurian Uplands had been raised the Devonian and Old Red Sandstone strata began to be deposited unconformably in inland seas and lakes bordering on these uplands—unconformably, because the strata of this mountainous surface had been contorted and worn down before the Old Red Sandstone was deposited upon it. It was thus that one formation, raised into dry land, supplied the materials for the next and others in succession. As the Ordovician and Silurian are therefore older than the Old Red Sandstone and the systems that followed it, a great gap exists in the geological history of Peebles and Selkirk up till the glacial epoch, deposits of which they have in abundance.

Graptolites from the Hartfell Shales, Mount Benger Burn, Yarrow, Selkirkshire

(1 *Diplograptus foliaceus*, 2 *Climacograptus bicornis*)

The fossils characteristic of the Ordovician and Silurian systems are called graptolites from their resemblance to a quill pen. They belong to the order of Hydrozoa. In the Silurian (Upper) the graptolites are nearly all single forms, as, for example, the *monograptus*. Branched forms as the *Didymograptus* and *Diplograptus* are very common in the Ordovician, but quite unknown in the Silurian system. Not only are systems distinguished by their characteristic fossils, but the sub-divisions or groups of systems are themselves distinguished in a similar manner. There are three places in the south-western borders of Peebles and Selkirk where fossils found in black shaly formations could not be identified with the fossils of the Silurian rocks found in the other parts as at Galashiels, where Professor Lapworth first discovered graptolites, and as at Grieston, where Nicol found many specimens of the *monograptus*. These places were Birkhill, Hart Fell and Glenkiln. Two of these groups were identified by means of their fossils with the groups of the Lower Silurian in Wales and the other with the group immediately above it and therefore as belonging to the Upper Silurian.

The district in Selkirkshire where the outcrops of Caradoc, Llandovery and Tarannon rocks (known as the "Ettrick Band") may best be observed, extends from Craigmichan Scaurs on the south-west of Capel Fell to Berry Bush in Tushielaw Burn, and is bounded on the north-west by the Yarrow and on the south-east by the Ettrick: an area of fifteen miles long by two miles broad and having upwards of fifty exposures. The line of separation between the Upper Silurian to the south-east and the Lower Silurian or Ordovician to the north-west follows the Kingledoors Burn to the Tweed, passes north of Dawyck, west of Stobo, to the Lyne, crossing the Tweed at its junction with that tributary. Passing north of Peebles over Hamilton Hill behind Neidpath it extends

Graptolite (*Monograptus Sedgwicki*)
from Grieston Quarry, Peeblesshire

Graptolites (*Monograptus Griestonensis*)
from Grieston Quarry, Peeblesshire

along the southern slopes of Makeness Kipps, where, making a return to form a lense-shaped bay, through which flows Leithen Water, it strikes north across the highroad between Innerleithen and Gorebridge, crosses the Gala at Crookston and cuts through the Lammermuirs to Whittinghame. North of the Ordovician area, the rest of Peeblesshire lying north-west of the line of fault (which practically follows the highway from Leadburn to Skirling) including the upper portion of the Lyne valley, belongs to the Old Red Sandstone formation.

Within the Silurian area a thin zone of limestone runs across the valley of the Tweed from Winkston by Drummelzier south-west by Wrae and reappears a little further on at Glencotho. What is perhaps a continuation of this limestone appears at Kilbucho.

Igneous rocks, usually consisting of porphyries, syenites, felstones and dolerites appear in dykes—*i.e.* vertical walls of igneous rock—coincident as a rule with the direction of the prevailing strike. The most prominent example of felsite porphyry is a group of dykes near Innerleithen on Priesthope Hill, the largest of which extends from above St Ronan's Mill to beyond Grieston Quarry for about 3½ miles. A section is exposed at Walkerburn. A series of outcrops of lava of Arenig age, the oldest exposed rock in the Southern Uplands, beginning beyond Biggar stretches in echelon order along the line of fault as far as Lamancha. These Arenig lavas form the base of the Southern Uplands and would be found anywhere in the region if one could bore deep enough. They appear in the Southern Uplands because the oldest

Silurian rocks have been upheaved at intervals all the way across from Ballantrae to the north of Peebles. The base of the Arenig lavas has, however, never been observed.

Many traces of glacial action and glacial drift occur in Peebles and Selkirk, the most important being the boulder-clay (*i.e.* the clay mixed with boulder stones deposited by the ice-sheet during the Glacial Period), the upper portion of which is often rudely stratified. The lower boulder-clay was mostly swept out of the valleys by the second glacier of this region, which left deposits of boulder-clay thickest in the valleys, but it is to be found up to a height of 1700 feet. It forms sloping shelves or terraces more or less denuded. Examples of these terraces, plateaux or banks, are to be found at Tweedshaws, at Lyne, in the Leithen valley, where also lower boulder-clay has been exposed with interbedded sands and gravels, at Glendean in the Quair valley and at Ettrick Toad Holes. Flutings, or markings due to glacial action on the hill slopes and valleys, are to be seen at Cademuir near Peebles, Kingledoors, Mossfennan, Drummelzier, near which stands Tinnis Castle, surrounded by a fragmentary ravine parallel to the river. In Drummelzier Burn on the slope of Finglen Hill another fragment of a water course seems to mark the bed of the stream which flowed to Tinnis Castle. At Cardrona, Traquair, and in Yarrow, these hollows or trenches of old water courses are also to be found. Terraces formed of banks of sand or gravel drift (left by glacial streams), called "kames," are to be seen in Lyne, at Sheriffmuir near

Lyne, in the Meldon valley, and at West Linton. Moraines (deposits left by glaciers) occur at Holylee, where the highway cuts through a terminal moraine, and in Manor, where a very striking series of moraines—one primary and several secondary—form a noticeable feature. In the same valley there is a *roche moutonnée*, round which the glacier cut its way so deeply that the engineers of the Edinburgh Water Trust failed to find a bottom. There are also moraines near Selkirk, and one, a fine example, on the road to Corbie Linn. Erratic blocks transported by glaciers are not found at a greater elevation than 1100 feet, but they are numerous in the upper grounds of Peebles and Selkirk.

The age and comparative softness of the rocks, the long denudation to which they have been subjected, have produced a striking absence of rugged masses. Another effect of glacial action not so noticeable, perhaps, but worth noting as a confirmation of the trend of the Tweed glacier, is that the western and south-western sides of the hills are always barer and steeper than the opposites sides, due to the forces of glacial action by which formation of "crag and tail" is produced.

7. Natural History.

The Southern Uplands from their inland and elevated situation and the uniformity of their physical features have a somewhat limited range of flora, while those plants of the Alpine series that are found are classified as sub-Alpine. Such are scurvy grass, the white cloud-berry in the black peat mosses of the Moorfoots, Yarrow and Ettrick, the yellow, starry and mossy saxifrages, the marsh thistle, monk's rhubarb, Alpine sedge, butterwort, *Festuca vivipara*, Alpine club-moss and the *Trientalis Europea*.

The hill pastures like the slopes of Cademuir gleam with the tiny starry-eyed *Helianthemum*, often with the yellow pansy as its neighbour. Further down, amongst the purple "sclidders," or by the drystone dykes, the pink foxglove shines vividly, sometimes amid masses of yellow broom. In summer the wild roses and the hawthorn, white and pink in summer, red as fire in winter, fringe the roadway. In the quieter meadows where the hills recede, or the current flows more gently, or in the dark marshy pools of the woods, as at Rachan or Soonhope, one comes upon the water forget-me-not or the rosebay willow herb, white grass of Parnassus, the marsh valerian, the queen of the meadow, and, very common in Tweed valley, the water-crowfoot. In the woods, the primrose, the wood-sorrel, the wood-anemone, and sometimes the harebell, where the canopy is thin, may be seen in leaf or flower.

Heather or ling is not uncommon, and white heather is found at Cademuir, Horsburgh, and Crookston. With the heather the red whortleberry (or Idaean vine) and the blaeberry (or bilberry,—not so common as else-where), are found on the heights; while the barberry, green and gold in summer, and green and scarlet in autumn, adorns the high hedges at Peebles, Linton,

Rachan, and Yarrow. Cotton grass, called when young, mosscrop, and when older, ling, is found in Ettrick and makes white in summer the heathery tracts at Leadburn; white bent, flying bent, stool bent, are all common on the hills in Yarrow and Ettrick. The bracken on the hill slopes and the curled rock brake are abundant. Hart's tongue and maidenhair fern are rare. The filmy fern is found in Megget.

The trees that grew in the Ettrick Forest were the birch, the Scots fir, the oak, the mountain ash, the alder, the ash, the elm, the hazel. Those introduced are the sycamore or plane tree, the larch, the spruce, and the silver fir (eighteenth century). The "Fauldshope Oaks," the largest clump of natural-grown oak in Selkirkshire, are small, gnarled, stunted trees, quite unlike the lofty trees for which the Forest was famed. Some years ago 300 acres of the south slope of Bowhill were enclosed to see if the indigenous trees would grow up. With few exceptions all the trees that grew up were natives. The oak, however, did not grow. The lessons that have been drawn from this, the "Howbottom Experiment," are that the old forest of Ettrick was not a stately and uniform growth of timber; and that the valleys were clothed with dense brushwood of hawthorn, birch and sallow, while on the hillsides and above the lower growth grew tall and noble trees of Scots fir, ash and oak.

The excavations at Newstead and the discoveries of remains in peat mosses show that the elk, the red deer, the roe, the wild boar, the fox, the badger, the wolf, and the hare must have been more or less numerous in the area in the period of the Roman Invasion. The horse was then represented by the forest pony (like the Shetland) and the Celtic pony (like that of Exmoor). There were two types of sheep, one with nearly upright, the other

Scots Pine

(*Edston Farm, near Peebles*)

with large, curved horns. Goats, apparently less common then than sheep, are still found wild in Megget, and near Hart Fell and Broad Law. The oxen of those times apparently belonged to the small Celtic shorthorn species, dark brown or black in colour with a red band on the back. Remains of the urus or wild ox were found at Lindean Loch in 1852, and at Whitmuir and Kerscleugh in Selkirkshire. But the names of the hills and valleys are adequate proof that the district was once haunted by these wild animals. In Ettrick Forest occur such names as Fawn's Law, Brock (Badger's) Hill, Earnsheugh (Eagle's Cliff), Deer Law, Bear Craig, Wolfhope, Buckscleuch, Swinebrae, Oxcleugh, Hartleap, Hyndhope, Gledcleugh (Hawkcliff).

In 1850 Sir John Hay introduced the fallow-deer to the woods at Eshiels; and the roe-deer found in the woods at Portmore and Dawyck is slowly working its way down to the wooded areas of Ettrick and Selkirk. The hill fox, often larger and greyer than in the lowlands, is still dreaded by the shepherds for their lambs and by the keepers for their pheasants. The brown rat, reported to have been first seen in 1777, spread through Peebles to Newlands by 1792; and by 1845 the black rat had disappeared from Manor. Plagues of voles (the short-tailed field vole) were so common from 1891 to 1893 in the south-west of Scotland including the west of Selkirkshire and Peeblesshire that a Royal Commission was appointed to deal with them; but before its report was ready, the voles were exterminated by the buzzards and owls, the tawny and the short-eared, which had collected in the district in great numbers. The brown hare is common in the fields, and the variable, blue or Alpine hare has spread over the whole area and beyond it since 1846, when it was introduced to the Manor district. The squirrel, at one time indigenous in the south, retired to the north on the destruction of the ancient woods and forests. In 1772 the Duchess of Buccleuch introduced it at Dalkeith, whence it has spread all over the Tweed area. It is specially destructive to young fir shoots. The otter, though becoming rarer, is still hunted in the Tweed, Yarrow, and Ettrick.

Peebles and Selkirk have not so many varieties of birds as other counties, for there is no sea-coast, and most of the area stands from 200 to 2000 above sea-level, and is largely moorland. There are, notwithstanding, about 100 species resident or migrant within the counties.

The thrush and the blackbird are plentiful. In the hills the ring-ouzel, or hill blackbird, though nowhere resident, takes the place of the merle. The whinchat has markedly increased, mainly in Yarrow and Ettrick, since 1904–5. The blackcap and garden-warbler are found in Ettrick, Yarrow and Tweed; while the sedge-warbler, the "Scottish nightingale," has been decreasing of late. But the chiff-chaff, the willow-wren, and the redstart are fairly common summer visitors. Of the wagtail family, the grey, the yellow, and the tree pippit are known, the third being numerous in the Ettrick, Yarrow, and Peebles hills, the second rare, having been last seen at Tushielaw in 1889. There has been a decrease of late years in the number of swallows. They used to be plentiful in Manor vale, where the cuckoo, also plentiful, drove them out of their nests. Of the finches the commonest is the chaffinch (Scots "Shilfa"); but the linnet (whinlintie), less common since the days of advanced farming, and the goldfinch are not unknown. The cross-bill has been seen at irregular intervals. The two buntings, the yellow-hammer and the red bunting, are common, while the snow bunting is a winter visitor in Ettrick valley, and also at Stobo and West Linton. The

1 Kingfisher, 2 Little Auk, 3, 4 Stormy Petrels

(*All shot in Selkirk*)

raven family breeds among the crags in Manor, Megget, and St Mary's; the carrion crow in Dawyck Woods; and the hooded or grey crow, locally confounded with the carrion crow, near St Mary's Loch. The jay has been increasing since 1897, but the magpie and the skylark have decreased in numbers. The night-jar, or goat-milker, wrongly persecuted by the keepers, the great spotted woodpecker, and the kingfisher are not unknown, and a specimen of the hoopoe was killed at Edston near Peebles in 1893.

Of birds of prey the owl is common in the area, the white or barn owl at Newark, Manor and Stobo, the long-eared owl in Ettrick. The tawny owl sometimes makes its nests in the trees in the woods. In the twelfth century high trees were left in Ettrick Forest for breeding places for the falcons. The peregrine falcon still breeds in the Traquair, Manor, and Tweedsmuir districts. A golden eagle was killed at Gameshope in 1833. An immature specimen of the osprey was shot at Cardrona in 1910. The buzzard may be seen every autumn in Peeblesshire. A rough-legged buzzard was shot at the Glen in 1876 and one at Eshiels in 1910. Five years ago the honey-buzzard was seen at Dawyck. The heron is common in the Tweed valley, and heronries were, or still are, to be found at the Haining, Cardrona, Portmore, Tweedsmuir, and St Mary's.

Geese are frequent in the region of the lochs in Selkirkshire, the commonest being the mallard, the golden eye, the shoveller and the tufted duck, the two latter in increasing numbers of late. The game birds, black grouse and red grouse (muirfowl), the indigenous grouse of Scotland, are common. The pheasant, often hand-reared, is numerous in the valleys of the Tweed. Coveys of partridges are common by the roadside. The "mud-dwellers," the golden plover, the lapwing (peewit or pease-weet), the curlew (whaup) haunt the lonely moors and hills in summer. Others less frequently seen are the common and the green sandpiper, while still more rarely come the greenshanks, the redshanks, the grey phalarope, and the stormy petrel. The common and the herring gull haunt the towns near sewage-tainted streams and garbage heaps. Black-headed gulls have colonies at Whitemoss, Linton, the Haining, Kingside Loch, and several mosses between Selkirk and Melrose.

In the Tweed and its tributaries trout and salmon are caught. In the lochs are found trout, perch, pike, and eels; and in the stream which joins the Loch o' the Lowes and St Mary's "a curious fish" used to be caught in the seventeenth century called "red-waimbs" (red-bellies) with forked tail. They were never seen except between Allhallows and Martinmas. Pennant in his *Tour* (1769) tells how in visiting Moyhall in Inverness he found Moy Lake full of trout and char, called in Scots "Red Weems." Red-belly is a common dialectic term for the char.

8. Climate and Rainfall.

By climate we mean the prevailing weather of a country; by weather, the state and behaviour of the atmosphere. These depend mainly upon temperature; and temperature is determined by latitude, altitude, season, prevailing winds, and proximity to the sea. Bulk for bulk, warm air is lighter than colder air; while water vapour is twice as light as air. Hence dryness, as well as temperature, affects the weight of the atmosphere. Warm and dry air may therefore be heavier than colder air. Air in motion will also naturally exercise less pressure than stationary masses of air.

In an area of low pressure the wind flows outwards in great spirals with a direction contrary to that of the hands of a clock. Such a condition of low pressure is called a *cyclone*. Cyclones accompany, like eddies in a river, the great drift of westerly and south-westerly winds which are the prevailing winds in our islands. From barometric readings, therefore, collected from various quarters, it is possible to plot out regions of cyclonic disturbance and so to foretell changes and disturbances in the weather. So also a region in which the pressure is high will, generally speaking, be one towards which winds will move in the same direction as the hands of a clock. Such a condition of high atmospheric pressure is called an *anti-cyclone*.

The region where the pressure is greatest in the Northern Hemisphere is along latitude 35° N.; and it is this belt of high pressure that has most influence on the climate of Great Britain, and, therefore, of Peebles and Selkirk. From the region of high pressure streams of air flow northwards to the North Pole and southwards to the Equator. But owing to the rotation of the earth from west to east, the winds become south-west winds and north-west winds respectively. It is with the former that we are concerned. These south-west winds, or "variable westerlies," are the prevailing winds of Great Britain, and consequently of Peebles and Selkirk. Records of winds give the following percentages for west, south-west, and south winds in Selkirkshire: Tinnis, for 25 years, 53·4; Bowerhope, near St Mary's, for 10 years, 60·9; Thirlestane, for three years, 60·5; and in Peeblesshire, at Stobo Castle, for five years, 51·39.

Seeing that the "westerlies" blow from a region of high pressure to one of low pressure they are said to follow the fall of the barometric gradient. That is to say, the winds should cut the lines of equal pressure at right angles, but, owing to the earth's rotation the winds are deflected, and so they cut the isobars at an acute angle. Roughly speaking, therefore, the isobars coincide in direction with that of the prevailing winds. The most important point to notice in connexion with the isobars is that as they pass over the Irish sea and St George's Channel, they curve downwards, and, as they pass over land, they curve upwards, the curve increasing in proportion to the width of the passage over the sea, or over the land.

The pressure within the counties is greatest in May and June, mostly in the latter month, and least in October. The barometer over a period of 40 years has stood highest at Galashiels with an average of 29·953, compared with readings taken at North Esk, the Glen, Stobo, Bowhill. The following table gives the average barometric pressure for 40 years (1856–95) with the average rainfall for the 40 years (1871–1910). It will be seen that the pressure and temperature vary indirectly, and the rainfall directly as the elevation:—

Place	Height ab. S. L.	Yearly Average		
		Press.	Temp.	Rain
North Esk Reservoir	1150	29·871	43·4	39·76
The Glen	765	29·876	44·7	40·60
Stobo	600	29·874	45·3	38·03
Bowhill	548	29·875	45·0	33·97
Gala	416	29·877	45·6	33·53

Other causes than that of elevation may, of course, have determined these means, and the lower temperature of Bowhill is no doubt due to a more south-westerly exposure than Stobo; but the regularity of the variation is sufficiently striking.

Since the sun is the predominating influence which determines annual temperature, the isothermals—lines of equal temperature—will follow mainly an east and west course, and the temperature will decrease as we pass northwards. The average rate of decrease in Great Britain is one degree for every 116 geographical miles. The "westerlies" bring heat and moisture to our shores, and, without the influence of the surrounding sea and these warm south-west winds, the climate of Great Britain would be so extreme that in January the temperature of Peebles would be equal to that of Greenland, or, in other words, drop 20°. Peebles and Selkirk being inland counties do not benefit to the same extent from these warm westerlies as the western seaboard counties. Edinburgh, although lying to the north, has a mean annual temperature 2° higher than that of Peebles and Selkirk, due to the proximity of Edinburgh to the sea; and to the greater elevation of Peebles and Selkirk, the temperature falling, on an average, 1° for every 270 feet of elevation. Within the counties themselves the variations in temperature depend mainly upon elevation and situation as regards the "westerlies." The highest stations will be the coldest, and the most westerly, other conditions remaining the same, the warmest.

The average annual rainfall of the British Isles is about 39½ inches. The driest part of the year in Scotland is generally April. The heaviest period of rainfall in Scotland is more irregular, occurring sometimes in winter and sometimes in summer. In Peebles and Selkirk, taking the results of 26 stations in 1909, we find that 14 places had their lowest rainfall in November. In 1910, out of 28 stations, all had their lowest rainfall in September. In 1909, out of 26 stations, 22 had their greatest rainfall in October. In 1910, out of 28, 17 had their greatest rainfall in August. North Esk reservoir with a record of 40 years gives a mean rainfall of 39·76 inches. The Glen for 20 years gives 40·60 inches; and the stations on the Talla catchment area for 15 years give from 62·70 at Talla Linns Foot up to 75·17 inches at Gameshope Farm. The highest mean fall in Selkirkshire is Borthwick Brae, with 44·29. The map shows very clearly that the average rainfall increases with altitude and with degree of exposure to the "westerlies." But the influence of position with respect to hills is greater than that of altitude.

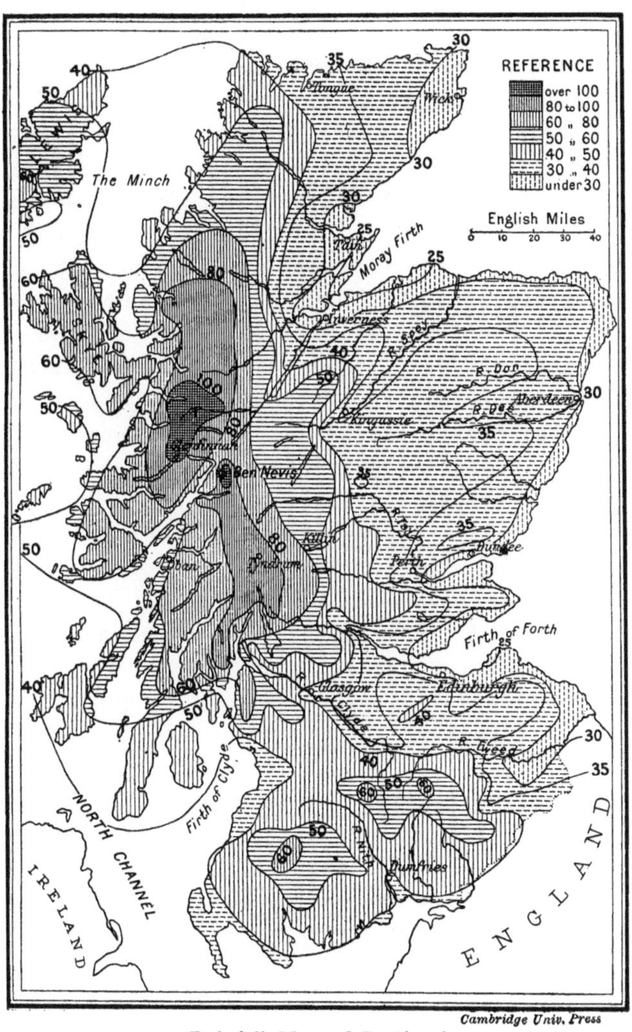

Rainfall Map of Scotland

(By Andrew Watt, M.A.)

In Peeblesshire and Selkirkshire the hills in the 60-inch zone are the highest in the Southern Uplands. The whole south-western portion of Selkirk, including Ettrick village and St Mary's Loch, lies within the 50-inch zone.

Peebles and Selkirk, therefore, have a less rainfall than the Western Highlands; but they have a greater rainfall than all the eastern counties of Scotland from Roxburgh to Sutherland. Most of the south of Scotland has a rainfall exceeding 40 inches, whereas roughly one-third of Scotland is embraced within the 30- to 40-inch zone. The crowding of the isohyets indicates a rapid change from one zone to another; and from the Grey Mare's Tail to Jedburgh, a distance of only 30 miles, we pass through five different zones of from 60 to 30 inches. As most of the river valleys run from south-west to north-east, the rain-bearing winds will bring moisture to both sides. Hence the hills are "the greenest that e'er the sun shone on." A Yarrow legend that the deluge came from the south-west, is no doubt due to the fact that all great rain storms and floods would come from that quarter.

9. People—Race. Language. Population.

Before and after the Roman invasion, successive waves of immigration passed over the Southern Uplands—Celtic Goidels, Celtic Brythons, Angles, Norsemen. The inhabitants prior to the first Celtic arrival are known as Iberians. Each wave of immigration influenced the population, and a striking result of this is seen in the place-names of Peebles and Selkirk. Gaelic and Cymric (*i.e.* British), English and Norse appear ; Gaelic rare, Cymric common, while, since some roots are the same in English and Norse, the Norse element has perhaps been under-estimated. Gaelic are *drum*, *cnoc*, *ra*, as in Drummelzier, Knockknowes, Rachan; Cymric are *caer*, *lin*, *pen*, *tor*, *tra*, *dre*, as in Cardrona, Linton, Lee Pen, Torwood, Traquair, Dreva; common to Gaelic and Cymric are *cad*, *loch*, *pol*, as in Caddon, Polmood. Most of the river-names are Cymric, as Tweed, Fruid, Talla, Manor, Leithen, Yarrow, Tima. Cymric names are remarkable for their melody, as is clear from the rhythm of the following couplet formed of place-names in order of locality :

> "Garlavin, Cardon, Cardrona, Caerlee,
> Penvenna, Penvalla, Trahenna, Traquair."

English roots are *ton*, *stead*, *cote*, *burgh*, *worth*, *heugh*, *law*, *edge*, *knowe*, *mount*, *head*: Norse are *grain* (a branching river or river valley), *scaur*, *myre*, *hope* (valley), *fell*, *rig* (hill), *holm*, *by*. Sometimes a name has elements with the same meaning from different tongues—a sign of mixture of peoples—as Knockknowes (Celtic and English), Venlawhill (Celtic and two layers of English). Norse words in common use, now or formerly, are *awns* (spikes of barley), *big* (build), *bygg* (barley), *gar*, *gimmer*, *leister*, *ling*, *lowe* (flame).

This district being for centuries part of the Anglian kingdom of Northumbria, its language is descended from that form of Northern English which came to be known as Lowland Scots. While many linguistic features are common to Peebles and Selkirk, each shire has certain peculiarities of its own, which tend more and more to disappear. The Selkirk speech, however, is the more distinctive. The reason apparently is that Ettrick and Yarrow districts owing to their geographical situation were less affected by the speech of the Scottish Court, and therefore by English and French influences, than Peebles. Peebles belongs to the dialect division known as Eastern Mid-Lowland, and Selkirk to that known as South Lowland.

The Selkirk dialect, probably the most direct descendant of the old Anglian speech, is characterised by a great variety of diphthongs and by its softness and flexibility of intonation. The distinctions are as follows: final *u* tends to become a diphthong. Peeblesshire *coo* in Selkirk is nearer *cuw* or English *cow*. Words like *see*, *me*, *we*, *he*, *dee* (die) become *sey*, *mey*, *wey*, etc. Peebles "you an' me 'll poo a pea" becomes in Selkirk "yow an' mey 'll puy a pey." Words like *bore* and *foal* are diphthongized into *buore* and *fuol*; words like *name*, *dale*, *tale* are pronounced *neh-um*, *deh-ul*, *teh-ul*. When the diphthongs *uo* (or long vowel *o*) and *ea* occur at the beginning of a word or are preceded by *h*, the first develops into *wu* and the second into *ye*. *Orchard* is *wurtshet*; *hole* is *hwull*; *whole* is *hyel*; *oats* is *yetts*; *one* is *yin*; *earl* is *yerl*; *home* is *hyem*; *sky* is *skyi*; *sword* is pronounced with the *w*. Finally the South Lowland is distinguished by its broad pronunciation of the vowel in *men*, which sounds like *a* in *man.* *Penny* is thus pronounced like *panny*, while *a* as in *battle* is often pronounced as *o* in *bottle*: even educated persons sometimes pronounce *a* in English *father* as *fother*.

The total population of Scotland at the last census was 4,759,445, 2,307,603 males, and 2,451,842 females,

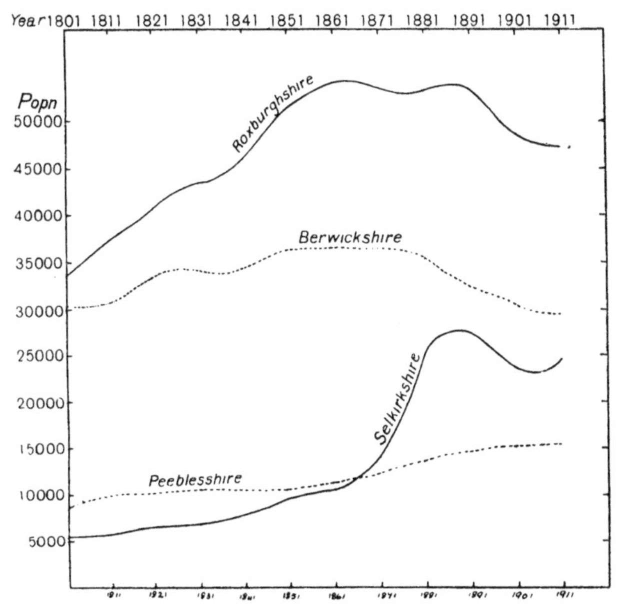

Curves showing the comparative growth of the populations of Peebles, Selkirk, Berwick and Roxburgh Shires

or 106·2 females to 100 males. The figures for Peebles are: males 7067, females 8191, total 15,258; or 114·4 females to 100 males; and for Selkirk: males 11,332, females 13,268, or 117·08 females to 100 males. Peebles has 43·93 persons to the square mile. Only five counties have a less density. Selkirk has 91·82 persons to the square mile; and eighteen counties have a less density. The increase of the population within the last 100 years has been greatest in the case of Selkirk. This is due to the fact that it was at Galashiels and Selkirk that the Tweed industry had its origin, reaching its greatest development between 1861 and 1881.

Peebles occupies a medium position between a rural and practically non-industrial county like Berwick, and an industrial district like that of Selkirk or of Roxburgh, the one with the busy manufacturing town of Galashiels, the other with that of Hawick.

10. Agriculture.

In the latter part of the eighteenth century a period of agricultural improvement began throughout Scotland. In our two counties improved methods of arable farming rapidly developed in the West Linton district; and further down the Tweed enterprising farmers ploughed land on hillsides which it would have been better to keep in pasture. Sheep farming also felt the impetus; and about 1785 the Cheviot sheep introduced on the hills of Peebles and Selkirk began to oust the Black-faced breed, while about 1845 a tremendous impulse was given to sheep farming in the district by the great development of the Tweed trade at Galashiels, Selkirk and Hawick. Since that time the tendency on the whole has been to withdraw land from arable farming and turn it into pasture, and for small holdings to disappear.

The following table gives the areas devoted to various purposes, with the percentage that each area bears to the whole.

	Peebles		Selkirk	
Total land area	acres 222,240	Percentage to total land area	acres 170,793	Percentage to total land area
Arable	27,500	12·37	16,000	9·36
Permanent grass	21,000	9·44	12,300	7·20
Mountain and heath for pasture	159,000	71·54	133,700	78·3
Woodland	11,300	5·10	5,200	3·04
Land otherwise occupied	3,440	1·55	3,593	2·10
Total	222,240	100·0	170,793	100·0

Arable land has been ploughed as far up as 900 to 1000 feet and wheat has been grown in Selkirkshire at a height of 700 feet. Since 1834 the area under the plough in Peebleshire has decreased, that in Selkirkshire increased, while in both the area under wood has been practically doubled.

The common rotation for crops in the counties is (1) corn (oats), (2) turnips or potatoes, (3) oats (or barley) sown with grass, (4) (5) (6) grass.

Owing, however, to its high elevation and moist climate the area is unsuited generally for the growth of cereals. But oats, turnips, grass and hay are readily grown. Wheat is practically unknown, while barley and potatoes are grown only to a trifling extent. Clover, sainfoin and rotation grasses are the largest crop in both counties: in Peebles 15,812 acres, in Selkirk, 8335. The total product of hay of all kinds for 1911 was in Peebles 6457 tons, the acreage being 5017, in Selkirk 3211 tons, the acreage being 3013; in each case the proportion of natural to artificial hay was about one half. The connexion between these cultivations and sheep farming is apparent; they can all be utilized for feeding purposes. Mixed farming, however, is supposed to be more economical for the simple reason that what is lost in the one department may be made good in the other. But the principal farming industry is sheep-rearing. Hill farmers breed to sell lambs; farmers lower down, while doing the same, also buy lambs for feeding purposes to sell in winter or spring.

In the time of James IV the total number of sheep in Ettrick Forest was 10,000—an extraordinary number it was then considered to be. But the Forest now bears eighteen times as many, the numbers for 1912 being:

Sheep	Peebles	Selkirk
Ewes breeding	89,427	81,259
Other sheep one year and over	23,662	19,343
Under one year	83,141	75,436
Total	196,230	176,038

About eighty years ago (1832) a fair estimate for Peebles would be 102,000, for Selkirk seventy to eighty thousand, or less than half of the present number.

Female sheep, from six to eighteen months old, kept for breeding, are called *hogs*; the next year *gimmers*; the fourth season *young ewes;* the fifth, and thereafter, *old ewes*; the males for fattening are called *wedders*; the others *tups* or *rams*.

The "Black-faced," "Tweed-dale," or "Forest" breed are horned, with black faces, black legs and coarse wool; compact, short legged, round bodied with rising forehead, and "kindly" feeders, that is, taking kindly to their pasture. The Cheviot breed was introduced in 1785 as the best adapted of the fine-woolled sheep for high, bleak situations. Hogg, "the Ettrick shepherd," fiercely opposed their introduction, lamenting that the black-faced "ewie wi' the crookit horn" should be banished from its native hills for those "white-faced gentry." Its introduction led to the planting of firwoods and the building of "stells" for shelter: noticeable features in the pastoral farms of the district. But in Peeblesshire, since 1864, owing to the losses of 1859–60, the Black-faced variety has been reverted to, the proportion in Peeblesshire now being three to two. In Selkirkshire, however, the sheep above one year are in the proportion of two-thirds Cheviots, one-quarter Black-faced, and the remainder Half-breds.

Before the days of sheep dip the wool had to be "smeared" or "salved" with tar[1] and butter. Farmers who advocated other methods were characterized as

[1] Sir Walter Scott had only one song, it was said, in his répertoire:

"Tarry 'oo is ill to spin."

This he used to sing at the Selkirkshire Pastoralists' Association.

"ignorant, inexperienced and revolutionary reforming farmers." Sheep farmers are now bound by the Regulations of the Board of Agriculture to have all their sheep dipped twice a year within certain specified dates.

Sheep are not shorn of their fleece till they are sixteen months old, and thereafter they are shorn every year,

Sheep-shearing at Henderland Farm, Megget

generally in July. The washing generally takes place from five to six days before the shearing, but as a rule the black faces are not washed. Their wool is sold "in the grease," in which condition it is said to keep better in transit, and the grease in the wool is manufactured into the by-product called "lanoline." The fleeces must be carefully tied up and all refuse kept out of the wool.

Cheviot wool is rolled up with the inside of the fleece outwards, and black-faced wool with the outside out. Hog wool is more valued than wedder wool.

The "clip," of course, varies. But in 1905 the average weight for Peeblesshire was $4\frac{1}{2}$ lbs. for ewes, and for other sheep $5\frac{1}{4}$ lbs.; for Selkirk 4 lbs. for ewes, and $4\frac{3}{4}$ lbs. for other sheep. The difference in weight between a washed and an unwashed fleece varies from 1 lb. to $1\frac{1}{2}$ lb., while the washed black-faced fleece is lighter than that of the Cheviot.

Sheep are subject to certain diseases, the most prevalent being "Braxy" and the "Louping Ill"; the former a species of inflammation, the latter of paralysis. The season for braxy is November to February, and in Peebles, Selkirk and Roxburgh the mortality from this disease sometimes reaches 25 per cent. The districts most affected are the hilly regions in the heart and in the south-west of Peeblesshire, a stretch of hilly country on the boundary line between Peebles and Selkirk and also stretching south-eastwards along the boundary line between Selkirk and Roxburgh.

The heather on sheep farms is burned once in nine years and new heather is ready to eat in three or four years; if the ground is mossy it may be in two years. Young heather is best both for farmer and sportsman. For long heather is of no use for cover unless the birds have also young heather to feed on. Hence some farmers contend that the proportion of young to long heather should be greater than it is. The dates for burning the heather are 10th December to 10th April, failing which application must be made to the landlord for special permission by the sheriff to have the time extended to the 25th April.

By 1714 Ettrick forest was completely denuded of its oaks. Then began an era of planting, which almost became a mania. Towards the close of the century, when Wordsworth with his sister Dorothy visited the district and found the

"Noble brotherhood of trees"

at Neidpath Castle cut down by the "Degenerate Douglas," they also found that a noticeable feature in the landscape was the raw new plantations surrounding a number of newly built mansion houses. The northern portion of Peeblesshire—containing the parishes of West Linton, Newlands, Eddleston, Lyne, Peebles and Traquair, with an area of 116,175 acres—has 6955¼ acres, or 6·0 per cent. under wood, while the parishes of Tweedsmuir, Broughton, Skirling, Kirkurd, Drummelzier, and Manor with an area of 106,424 acres have only 4370¾ acres or 4·1 per cent. under wood.

In Selkirkshire the parishes of Caddonfoot, Galashiels, Yarrow and Selkirk, amounting to 92,412 acres, have 3989¾ acres or 4·3 per cent. under wood, while the parishes of Ashkirk, Ettrick, and Kirkhope, containing 78,349 acres, have only 1303¾ acres or 1·6 per cent. under wood. Peebles is therefore nearly twice as well wooded as Selkirk, but is itself about three times less well wooded than the best-wooded districts of Scotland. Dawyck woods planted by Sir James Naesmyth, assisted it is said by Linnaeus, whose pupil he was, cover some

2800 acres and are amongst the most famous woods in the south of Scotland. Other well-known woods are to be found at Stobo, Haystoun, Bowhill, the Haining and Hangingshaw. The trees planted for economic purposes are mainly the Douglas pine (which is extensively planted),

Oldest Larch in Scotland

(*Planted at Kailzie by Sir James Naesmyth of Posso in* 1725)

the Scots fir, the clear pine, the larch, and the sycamore (Scots plane tree).

A special cultivation of interest is found in the vineries of Clovenfords. Established in 1868, the vineries and plant-houses cover nearly six acres and are heated by some six miles of pipes. They produce annually about 15,000 pounds of grapes, the best-flavoured being the Duke of Buccleuch, raised by the founder, who was the Duke's gardener. Tomatoes, cucumbers, melons, palms, araucarias, dracaenas and aspidistras are grown as well as grapes.

11. The Manufacture of Wool.

In a district famous for sheep, the chief manufacture is naturally that of wool. At one time Selkirk was famous for its shoemaking. The "Souters," however, with their "single-soled shoon" have long since disappeared. "Single-soled shoon" were brogues with a single thin sole, the purchaser himself sewing on another of thick leather. "Souter" has continued to be the distinctive appellation of the inhabitants of Selkirk. The quaint ceremony of "licking the birse" is still performed by the recipient of the honorary freedom of the Burgh, the "birse" being the bristles with which shoemakers point their "lingles" or thread, and the licking being performed by dipping the bunch in wine and then drawing it through the lips.

In 1587 Parliament passed an Act to encourage the settlement of Flemish craftsmen and the employment of Scottish apprentices. About this time, also, we find the first mention of the manufacture of wool at Galashiels, which then had two "wauk" mills. By the seventeenth century three mills were busy felting or milling the webs made from the wool of the district and spun by the women in their houses. The thieves of Liddesdale held the

Galashiels "hodden grey" in high repute. During the days of the Civil War numerous acts were passed to encourage woollen manufacture in Scotland. The Board of Manufactures in 1728 appointed in Galashiels, Hawick, Jedburgh, Peebles, and Lauder, persons skilled in sorting, stapling and washing coarse, tarred wool. Each received a salary of £20 and also utensils. These grants were continued to the woollen trade till 1840. In 1835 Galashiels manufacturers built mills in Selkirk; about 1850 the first cloth-mill was established in Peebles; and thereafter the trade took root in Innerleithen and Walkerburn.

At the beginning of the nineteenth century the kind of cloth manufactured was shepherd tartan, of which travelling cloaks were made. Trousers were made of the same pattern, and Sir Walter Scott's may still be seen at Abbotsford. Mr Dickson of Peebles manufactured trousers of the plaid pattern for the London market, and the only variation of pattern attempted was the size of the black and white check. Then checks of black and brown were introduced and other colours tried. Following the checks, twills were tried, and new combinations of colours followed. Every change gave the trade a fresh impetus, and Scottish fancy woollens became the fashion. The local supply of wool proved inadequate, even though a corresponding development took place in pastoral farming; and in 1834 fine wool was imported from abroad. Within six years four-fifths of the wool was imported—at first the fine merino of the continent, but soon the more suitable wool of the colonies was employed.

From the 400,000 sheep in the district the yield of unwashed wool is upwards of 2,000,000 lbs. As the district probably possesses more sheep per acre than any other region in the world, it is not difficult to understand why the Scotch Tweed trade should find its home in the valleys of the Tweed and its tributaries. But, great though the home supply is, it is insufficient to meet more than one-tenth of the trade requirements.

There are 43 woollen mills, using annually about 18 million lbs. of raw wool, valued at over £1,000,000 sterling. These mills contain 200 sets of carding machines, about 160,000 mule spindles, and 1900 power looms, employing altogether about 7500 workpeople, earning, it is estimated, £375,000 in wages per annum. The capital sunk in the woollen industry of the two counties will exceed two millions sterling. Fully 60 per cent. of the Scotch Tweed produced is manufactured in the counties of Peebles and Selkirk.

The Scotch Tweed manufacturers have always been strong supporters of technical education. In 1883 classes for instruction in the technique of manufacture were commenced in Galashiels under the auspices of the Manufacturers' Corporation. In later years the classes attained a remarkable degree of success and their good work was so appreciated that, when the manufacturers were invited to contribute towards a scheme for a Technical College for the south of Scotland, a sum of £11,000 was readily forthcoming, which, augmented by an equivalent grant from Government, enabled the promoters to erect a college worthy of the traditions and importance of the

Power Looms

Warping Machines

(*March Street Mills, Peebles*)

woollen trade. Galashiels has become a name to conjure with throughout the world not only on account of the excellence of its "Tweed," but also on account of the skill of its Tweed designers, and in consequence many Borderers are to be found all over England, Ireland,

Technical College, Galashiels

Europe, America, and the colonies holding high positions in woollen mills.

The kinds of cloth manufactured in Galashiels, Selkirk, and Peebles vary from time to time, and it may happen that while trade is busy in one town or in one manufactory of a town, it is extremely slack in another town or factory. The staple manufacture of the district, however, is Cheviot cloths suitable for sport and motoring and out-of-doors wear, Saxony and worsteds not lending themselves to the make-up of garments for such purposes. It will be seen, therefore, that in the Tweed manufacture a great deal depends upon the readiness with which the manufacturer can anticipate and supply the popular taste.

The origin of the word "Tweed" in its industrial sense is interesting. In the early part of the nineteenth century a considerable trade in Scotch "Tweels" had sprung up with London merchants. In 1826 a firm in the south of Scotland consigned a quantity of these goods to a leading woollen warehouseman in London. The invoice clerk by a slip transformed "Tweels" into "Tweeds"; and the merchant, thinking this an appropriate designation, repeated more "Tweeds." The name and cloth caught the public favour, and "Tweed" is now the accepted trade description throughout the world.

12. Minerals.

Except in north-west Peeblesshire, no rocks of economic value occur in the two counties; unless greywacke (whinstone), useful for building and for road-making, may be so regarded. Before the period of tree-planting, whinstone was much in evidence as stone-wall fences. The whinstone being a stratified rock splits readily with a clean fracture. It has undergone many contortions, which render it difficult to deal with for building purposes, but the stonemasons of the district are famous for their skill in its manipulation, producing as they do with only a hammer and trowel beautifully-faced walls. Freestone abounds in the carboniferous tracts, white and yellow as at Carlops, chocolate-coloured as at West Linton. In the Dod Wood at Kirkurd are numerous old and new quarries of white and red sandstone, where the red stone of the buildings at Lyne Camp were probably obtained. Previous to 1841, before the geological record was thoroughly understood, the carbonaceous shales of coal and limestone were wrought at Carlops; and not so long ago a coal pit was worked at Macbie Hill, where still a little mining is done. Attempts were also made to find coal at Lindean and Galashiels; and anthracite was said to have been got at Grieston and Caddonfoot. But these attempts were bound to fail, because the sandstones, the limestone, and the millstone grit of the West Linton district all lie beneath the coal measures, which are naturally thickest in the middle of their hollow basin, and thinnest at the upturned edges. Such coal as is found in the district will be "edge coal"; while "anthracite" found in Silurian strata is either black shale or has been formed from quantities of embedded animal matter.

Lead used to be worked on the Medwyn in the sixteenth century, and the excavations are now called "Silver Holes" from the fact that silver was once obtained there. In the seventeenth century a lead mine was said to exist on the north side of Selkirk, at the head of the Linglie Burn, and a silver mine at Windy Neil. Lead has also been mined for at the Bold Burn, at Grieston, at Windlestrae, at Kershope in Yarrow, and at Innerleithen, where smelting furnaces were discovered four feet beneath the surface in the churchyard. Gold is said to have been found in Henderland, in Glengaber, and Mount Benger Burns, in the reign of James V. A specimen from Glengaber Burn is preserved in the Peebles Museum. Gold is also recorded as having been obtained in the Douglas Braes at Douglas Craig and in Linglie Burn. The Regent Morton had a contract for working gold at Henderland. But the enterprise was unsuccessful. Veins of haematite occur here and there in Silurian rock. At Noble House a bed of red haematite shale lies among the green shales of the district, and was worked some twenty years ago. Iron pyrites occur at Bowerhope, and oxide of iron is found in many of the mosses. Silurian shales have often been worked for slate, as at Stobo and Grieston quarries. Out of the former many of the houses in old Edinburgh are said to have been roofed. These quarries are no longer worked, either because they are exhausted, or because better material is now more easily obtained. The felsite near Innerleithen has been used for making curling stones. Lime quarries are common in the West Linton district; and lochs in Selkirk have sometimes been drained for their marl, a mixture of lime and clay, invaluable to the farmer.

Mineral springs are fairly numerous. A century ago the well-known chalybeate spring at Innerleithen made the village a fashionable summer-resort and furnished Sir Walter Scott with a setting for his romance, *St Ronan's*

Well. This spring used to be known as the "Doo Well" because of the pigeons that flocked to it. A sulphurous spring at Castlecraig had the reputation of being stronger than that of Moffat. At Rutherford near Carlops there is a chalybeate well, "Heavenly Aqua"; another, "Philip's Well," at Catslacknowe in Selkirk; and two at Bowerhope. Calcareous springs have been found in fifteen different places in Yarrow.

St Ronan's Well, Innerleithen

Alluvium peat is found in many of the hills, as is shown by the not uncommon designation of "Peat Law." The hills of Manor, the Moorfoots, and Auchencorth Moss, near Leadburn, are the best known districts for peat. Experiments were made in the compression of peat by the minister of Traquair about 1834; but, owing to railway extension and the cheapening in the price of coal, the digging of peat is now confined to the remoter districts amongst the hills.

13. Fishing.

For salmon, grilse or sea-trout few rivers can surpass the Tweed. Though not free from impurities near the manufacturing centres, it may on the whole be designated a clear, clean river. It is fairly free from rocks and overhanging woods, while its gravelly bottom with loose stones of moderate size, is suitable for spawning, and furnishes abundant and suitable feeding for the fish. The river, neither swift nor sullen, but with complete and uninterrupted charm for the angler, ripples in silvery streams from pool to pool.

Trout fishing, except near the towns, where it is overdone, is good; and salmon fishing in its season, from Peebles to Berwick, is excellent. Par and smolts are illegal capture till the first of June, and the close season lasts from October to January inclusive. Neither trout nor salmon fishing is quite so good as formerly—due no doubt to extensive drainage, causing the flood waters now to run off in days instead of in weeks; to poaching; and to fishing out of season. An Angling Improvement Association has been formed at Peebles to check the two latter evils; and certain proprietors in the district who proposed to close their waters have now leased them to the Association, which controls a stretch of water from

Manor Bridge to the march at Elibank, between Peebles and Selkirk. Throughout its 100 miles Tweed has 316 named Salmon casts; 55 casts from "Inch" three miles above Peebles to "Kameknowehead" near Elibank. The remaining 261 casts from "Kameknowehead" to "Low

Bend on the Tweed near Yair

Bells" near Berwick are either let, or in the hands of the proprietors.

The principal tributaries and sub-tributaries—most of them interesting and picturesque—in which good angling may be had, are Cor, Fruid, Gameshope, Hearthstone, Holms, Kingledoors, Menzion, Polmood, Stanhope, Talla, Lyne, Tarth, Manor, Quair. The Peebleshire Lochs are not of much account; but mention may be made of Portmore (pike, perch, trout), Gameshope, Slipperfield (pike and perch but no trout), Talla Reservoir, and North Esk Reservoir.

Yarrow, surpassing Tweed in poetical and romantic lore, approaches it in fishing fame. Beyond the rocks and trees, there are some fine casts; as Levinshope Burn to Deuchar Mill; from Sundhope for a mile up (the best angling part of Yarrow); Eldinhope Burn and the Douglas Burn, tributaries on the left. St Mary's Loch, an expansion of Yarrow, can be fished all round the shore. In this loch the trout are in the majority, but pike and perch are on the increase. In the Loch o' the Lowes there were no trout twenty years ago, but now there are a few, mainly on the south shore and superior in quality to those of St Mary's, while the pike as edible fish are superior to those taken elsewhere and often attain a great size. Kirkstead, Glengaber and Winterhope Burns are good trouting streams. The Ettrick is a salmon stream. But trout are hard to catch. The best angling part is from Tushielaw Inn to the foot of Tima, a distance of three miles, while its tributaries, particularly the Bailie Burn, the Rankleburn, the Tima, with Glenkerry, all give good sport. Of the Lochs other than St Mary's and the Loch o' the Lowes, the best are the Haining, Headshaw, five miles from Selkirk, Essenside, Alemuir, Hellmuir, the Shaws Lochs and Acremoor.

14. History of the Counties.

The inhabitants of Peebles and Selkirk are a mixture of many races, the process of whose consolidation did not terminate till a Scottish king sat upon the English throne. Hence one may assert that over the Southern Uplands the tide of war has ebbed and flowed for more than two thousand years.

It was David I who began to civilize the Borders. By the time of the Alexanders, Scotland, and particularly the Lowlands, had attained a high degree of civilization. The Wars of Succession, however, checked this for many years; and no part of the Lowlands suffered more than Peebles and Selkirk. The connexion of the shires with these wars is not unimportant. The men of the Forest fought under Wallace at Falkirk; and the noble and handsome forms of those who fell roused the pitying admiration of the English Chronicler of the fight. Wallace after his desertion by the nobles at Irvine took refuge in the Forest; and a Peeblesshire baron, Sir Simon Fraser the younger, the patriot's friend and companion-in-arms, and the hero of Roslin and of Methven, shared eventually Wallace's fate. The Good Sir James was lord of Ettrick Forest. The Knight of Liddesdale, slain by his kinsman near Broadmeadows, was one of the band of heroes who won back from the English the castles they had captured in the time of David II.

In the fourteenth century the Borders on both sides were divided into three Marches: East, West, and

Middle. Peebles and Selkirk were included in the Middle March. Over each March was set a Warden, and at stated intervals on days of truce Warden Courts were held. Thus grew up the Border Laws which dealt with fugitive serfs, and with offences committed by Borderers on either side of the boundary, such as manslaughter, and theft of goods or cattle. The first code of Border Laws was drawn up in 1249; the second exactly two hundred years after. They were revised from time to time till the Union, when they became null and void.

Flodden Memorial, Selkirk

Various attempts were made to establish order; notably by James II in his contest with the Black Douglas, whose territory in Ettrick Forest he more than once invaded and whom he finally crushed at Arkinholm in 1454. James IV also made at least one famous expedition to the Forest, when he exacted submission from the "Outlaw Murray." Flodden, which so greatly enriched the fame and traditions of the Forest, gave only a short respite to the state of anarchy to which the Burgh

Records of Peebles bear frequent and eloquent testimony. Brawls and fights in the streets, rapine, raid, and murder were the order of the day. The Tweedies of Drummelzier, the Scotts of Thirlestane, and other clans were neither "to haud nor to bind."

It was not, however, till the relentless persecution of Dacre after Flodden that life on the Borders was brought to a state of positive demoralization. "The Borderers," says Creighton, "ceased to regard themselves as bound by any laws save that of the family tie, and degenerated into gangs of brigands whose hand was against every man, and who made little distinction between friend and foe." Hence it is that James V is best known for his determined attempts to restore law and order upon the Borders. In 1529 he visited Peebles and Jedburgh for this purpose. The following year he resumed the task, and with a sufficient force followed the "Thief's Road" across the Tweed, up by the Lour, round the Scrape and Dollar Law, then down the Craigierig Burn to Henderland, where he arrested William Cockburn. From there he went to Tushielaw, where he surprised Adam Scott, "the King of the Borders." The two blackmailers were taken to Edinburgh and executed. *The Border Widow's Lament* commemorates the burial of Cockburn. But even these stern measures failed to awe the greater barons, whom James suspected of connivance at the depredations of their "kindly tenants." He, therefore, in the same year, caused several of them to be imprisoned. This alienated the Border barons; and James felt the bitter result of their defection at the rout of Solway Moss.

In the reign of Mary the war with England united the Borderers against their "Auld Enemy," and even Angus returned from exile to break a spear in defence of his country and the honour of his ancestors, whose tombs Latoun had defaced. The victory of Ancrum Moor roused Henry to fury, and the following year he dispatched Hertford to take vengeance on the Scots. The tale of his burnings and slaughterings is appalling. Peebles was burned to the ground with 250 towns and villages in the Tweed area besides towers and castles and monasteries. Three years after Henry's death came peace between England and Scotland; and the lawlessness of the Borders grew more rampant than ever. On Queen Mary's return from France, Moray was entrusted with the duty of restoring order. His policy—afterwards adopted by Morton and by James VI—was extermination. Yet one of the last Border raids—perhaps the most daring of all—was conducted in James's reign by the king's own Warden in 1596, when the "Bold Buccleuch" rescued "Kinmont Willie" from the castle of Carlisle. This deed, the fame of which resounded through Europe, nearly brought the two countries to war. It was about this time that the Border counties began to be known as the Middle Shires and a commission was appointed to establish order therein. Special courts were appointed in place of the old Warden Courts, at such places as Peebles, Hawick, and Jedburgh. Through the expeditious severity displayed by Dunbar the Commissioner at Jedburgh, "Jethart Justice" came to signify "hang first and try afterwards."

The Reformation had had little immediate effect upon the Borderers, nor did the constitutional and religious struggle of the seventeenth century strongly appeal to them. The enthusiasm for the Covenant was less ardent than in Galloway or Ayrshire, if an exception may be made for the west of Selkirkshire and for the Galashiels district. Yet when Montrose, seeking for support to the king, reached Kelso, he received little encouragement. Montrose advanced to Selkirk and took up his position at Philiphaugh. Leslie, receiving word of his proximity, marched with his main body on Selkirk, sending a force round Linglie hill to attack Montrose in the flank and the rear. At Leslie's unexpected attack, the royal troops fled in rout over the hills to the west and north. Douglas and Montrose, cutting their way through Leslie's lines, fled over Minchmoor, and reached Traquair House, where they were denied admittance. Making their way through the Tweed at Howford, they reached Peebles. From there, they escaped across to Clydesdale. In the year after his victory at Dunbar, Cromwell dispatched a force under Lambert to besiege Neidpath, held by the Earl of Tweeddale. The attack was made from the south side of the river and after a brave defence the Earl surrendered.

To the fiasco of the "Fifteen" Selkirk gave a supply of shoes and a contribution of £10. In 1745 the town of Selkirk, at the request of the city of Edinburgh, furnished the Pretender with 2000 pairs of shoes for his army. After Prestonpans, the Prince advanced towards England in two main divisions. The first column marched by Auchendinny to Peebles, thence to Broughton,

Tweedsmuir, and Moffat. At Peebles the contingent occupied the field west of Hay Lodge, and the town-mills were kept busy on the Sunday to supply the troops with meal. The main column, under the command of the Prince, went by Lauder and Kelso, whilst the baggage party went by Galashiels and Selkirk. Charles Edward is said to have visited Traquair; but the Earl declined to join his cause, and to soften his refusal, declared that the gates would remain closed till Charles Stewart re-entered them as Sovereign of the Kingdom.

The war with Napoleon aroused strong feelings of patriotism. The old fighting instinct asserted itself again, and Peeblesshire, after the Peace of Amiens, raised a levy of foot and horse which outnumbered per 1000 of the population that of any other county in Scotland. Nor was Selkirk less enthusiastic; for on the occasion of the "False Alarm" on the night of January 31st, 1804, the Borderers responded gallantly to the ancient signal of the Beacon Lights, and the Selkirkshire yeomanry made a notable march, reaching Dalkeith by one o'clock the following morning.

15. Antiquities—Pre-historic, British, Roman.

In pre-historic days, the Neolithic men buried their dead in long barrows or mounds, while the later Celts buried in round barrows. Long barrows contain no metal weapons; round barrows have bronze weapons and ornaments as well as stone. In the bronze age, gold ornaments are also found. The sepulchral cairn, however, is commoner in Peebles and Selkirk than the barrow. Tombs of the ancient Celts have been occasionally discovered in almost every parish in Peeblesshire; but most frequently in the west, especially in the Lyne valley.

The ancient Britons have also left numerous hill-forts,

Catrail Fort at Rink

their houses or defences, which existed before, during, or after the Roman occupation. No fewer than 83 of these hill-forts have been surveyed in Peeblesshire. They are most numerous in the west and north-west of the county, rare in Tweedsmuir, in the Quair and in the Leithen valleys, unknown on the slopes of the Pentlands and in the valley between these hills and the Southern Uplands.

In Selkirk they are not to be found in the middle valleys of Ettrick and Yarrow; and only nine in all exist in the eastern part of Selkirk, the most important being the Rink. The forts are usually situated, at an elevation ranging from 1000 to 1400 feet, on terminal spurs, as East Cademuir; on isolated hills, as Macbeth's Castle; on the slopes of valleys, as Harehope; or in the valley itself, as Stirkfield, Broughton. Two-thirds of them have been constructed entirely of stone, the rest of earth, or of earth and stone. Their general form is curvilinear, modified to suit the outline of the surface. But it is not possible to say whether the walls were built, or simply piled up. In the fort at Dreva, however, traces of building have been seen. Some forts, as Upper Cademuir, have treble rings; some, as Cardrona, double; and some, as East Cademuir, single rings. The circumference varies, roughly from 150 yards at East Cademuir to 600 yards at Upper Cademuir. Two stone forts, West Cademuir and Dreva, are defended by groups of stones at a lower level than the camp, forming a sort of *chevaux de frise*, a feature found nowhere else in Scotland.

None of these forts equals in interest that on Torwoodlee hill a few miles from Galashiels, 300 feet above Gala Water and situated within the area of a British camp on Crossleehill. It belongs to the type of fort known as a broch. Brochs are dry-built circular castles. They are characteristic of the Celtic area, outside of which they have never been found. They belong to post-Roman times; their relics are Celtic, Roman, and post-Roman. The remains of the Torwoodlee broch measure 75 feet, and the enclosed court 40 feet in diameter, the height of the walls being about three feet. The entrance passage is on the east side and must have been closed by a door. At the main entrance was a guard room within the thickness of the wall, and on the south-west side there are the remains of a staircase which would lead to the upper galleries of the tower, sometimes five or six in number, the floor of one forming the roof of the other. The broch of Torwoodlee is thus larger than that of Mousa. The relics of the brochs show that their occupants hunted in the forests; kept flocks and herds; cultivated grain; fished rivers and seas; and were acquainted with the arts of weaving and pottery, metal, wood, and stone work. The relics of Torwoodlee broch consist mainly of pottery, glass, enamels, and iron implements.

The broch of Torwoodlee seems to be the terminus of the Catrail, one of the most wonderful monuments of antiquity in the south of Scotland. It consists of a ditch with a double mound, one on each side, obliterated in many places, in others, distinct. Even where no trench or mound exists, its course can often be traced by the lighter shade of the grass, by the darker green of the young corn, or in winter, by the longer-lying snow. Its course, as it halves Selkirkshire in two, stretches over Tweed, Yarrow and Ettrick for 50 miles from Torwoodlee camp in the north-east of Selkirkshire to the slopes of Peel Fell in the Cheviots. Where it is perfect the width of the fosse from the summit of one mound to another, varies from 23½ feet to 18½ feet; the width of the

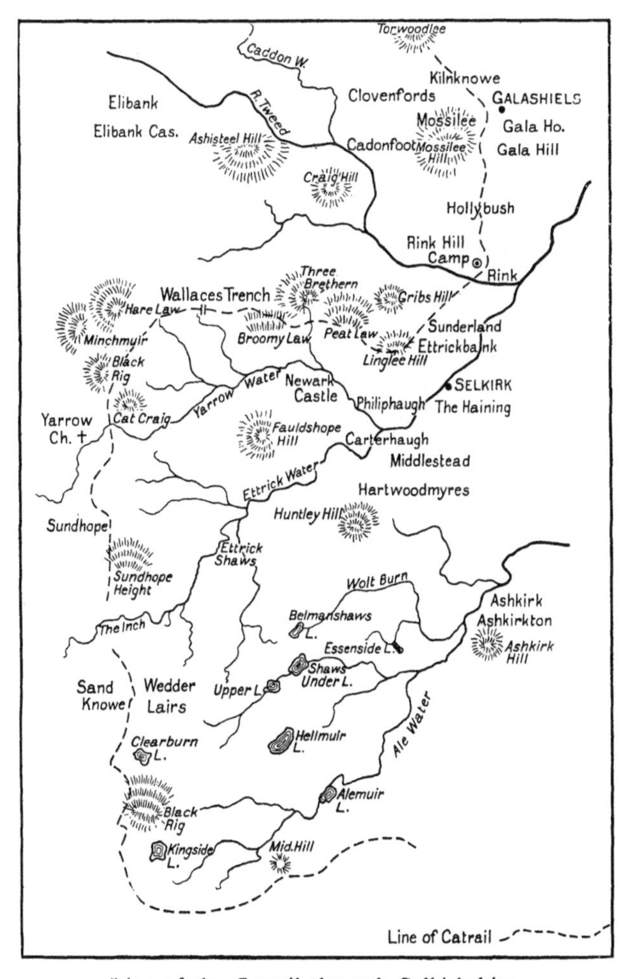

Line of the Catrail through Selkirkshire

bottom of the ditch is on an average six feet; and the distance from the summit of the slope to the bottom is 10 feet. Three theories have been advanced to explain the Catrail: (1) a line of defence by the Britons against the English; (2) a territorial boundary between Anglian-Bernicia on the east and British Cumbria on the west; (3) and best, a strategic road between the greater forts constructed by the Romanized Britons to check the English invasion.

Lyne Camp was a *castellum* or fortified camp, probably on a Roman road leading to Antonine's Wall. It is situated on the plateau of a moraine about 100 feet above Lyne Water, towards which it slopes on the west and south. The north and east sides of the camp were protected by a morass, the west and south by the river and its sloping banks, and the east by a natural mound, now covered with trees. Two annexes, one on the north-west angle, the other on the south-west, filled up the vacant spaces between the edge of the marsh on the north and the slope on the south sides. On the east, towards which the camp faced, there were three lines of defence 140 feet in width; on the south the breadth of the fortifications was reduced to 120 feet, on the north-east (where the mounds are most clearly marked) to 85 feet; on the north-west to 45 feet; while on the south-west there was only one rampart with its trench. The variation in the width of the defences was dependent, of course, on the amount of natural protection afforded by the slope or by the marsh. There were no gates or barricades on the east, but there were gates on the north

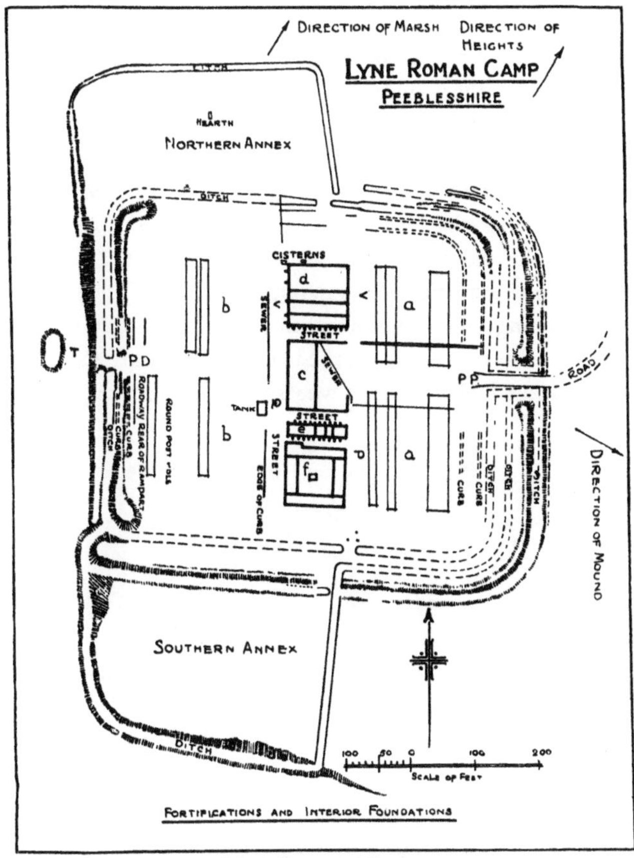

Lyne Roman Camp

Explanation of Plan: a, *a* the Pretentura, *b*, *b* the Retentura, *c*, *d*, *e*, *f* a line of 4 (probably 5) stone buildings, *c* the Praetorium or Principia, *d* the officers' quarter buttressed portion next to *c* probably a horreum, *e* a horreum, *f* officers' quarters, the small square a pit, *PP* Porta Praetoria, *PD* Porta Decumana, *V*, *P* the Via Principalis, *V*, *Q* the Via Quintana, the dots at the gateways represent postholes, *T* a traverse opposite western entrance. (Mr James Curle, author of *Newstead Fort and Camp*, thinks that *d* probably consist of two buildings.)

and south. The south entrance opened into the annex, from which there must have been a bridge. A short portion of a road remains leading north-east and then south-east from the eastern wall of the camp. In a pit in the courtyard of the annex to the south, were found the few relics that were discovered: some Samian ware, glass, nails, and two coins, a denarius of Titus (A.D. 79) and a brass sestertius of Trajan (A.D. 104–110).

Standing stones or megaliths, of great antiquity, are found in Manor (a cup-marked stone); at Lour, also cup-marked; at Dollar Law, Tweedsmuir, Sheriffmuir (Lyne),

Roman Coin found at Bellanrig in Manor, 1910

Obverse:—ANTONINVS AVG. PIVS P.P.TR.P. *i.e.* Antoninus Augustus Pius, Father of his country with Tribunician Power. (Date, probably 145.)

Reverse:—COS. IIII. *i.e.* the fourth year of his consulship. The letters, S.C. also occur.

Cardrona, "Warrior's Rest" (Yarrow). Some of these are no doubt monumental. The eleven stones, eight of which are standing and three lying down, on Blackhouse Heights, said to mark the "Douglas Tragedy," are according to Professor Veitch older than feudal times. The stones at "Warrior's Rest" were boldly, but without warrant, linked by Sir Walter Scott with the legend of the "Dowie Dens."

Flint arrows, stone axes and hammers, mostly of other stone than flint; bronze axes, flat, flanged, and socketed—the three stages of their evolution—have been found at various places, but mainly in the west. A food urn of rare and elegant design was found at Darnhall, a bronze caldron at Hattonknowe, a Roman patella at Stanhope, and gold ornaments at Shawhill.

16. Architecture—(*a*) Ecclesiastical.

The earliest church buildings in Scotland were usually of wood and clay, resting upon stone foundations. Church settlements of a very early date existed in Peeblesshire at Stobo, Kingledoors, Glenholm and Drummelzier. Kingledoors Chapel in Tweedsmuir was either founded by St Cuthbert or, like the last two, dedicated to him soon after his death in 687 A.D. Churches in the twelfth century existed at Peebles and Traquair; and, if Selkirk means "Kirk of the Shiels," in Ettrick Forest long before the twelfth century. But the remains of ancient churches within the shires are singularly rare and of little architectural interest.

The Church of St Andrew in Peebles was founded by Bishop Jocelin of Glasgow in 1195, in the reign of William the Lyon, and therefore belongs to the transition period of Norman to Early English. The walls were built of undressed whinstone; and a tall square tower, "restored" by Sir William Chambers, at the west end of what must have been a spacious building, is all that now remains of the structure. In 1406 it was burned by Umfraville, "Robin Mend the Market," and nearly one hundred and fifty years afterwards it suffered when Hertford destroyed the town by fire. At the Reformation

Tower of St Andrew's Parish Church, Peebles, before restoration

in 1560 it was abandoned; and there is a tradition that Lambert, when besieging Neidpath Castle, stabled his horses in the church, which by that time had fallen into ruins.

The Church of the Holy Cross was founded by

Alexander III in 1261. In that year, says John of Fordun, a cross was found at Peebles, and near the cross an urn, with the relics of the martyr St Nicholas, supposed to have been massacred in the reign of Diocletian. Crowds of people flocked to the spot, and many miracles were performed. More than 200 years after, in the reign of James II, a monastery was added to the church. The unusual position of the monastery on the north side of the church, Dr Gunn supposes to be due to the fact that the niche containing the relics of St Nicholas was on the south wall of the church. The space opposite this side of the church would naturally be the resort of the crowds of pilgrims who resorted thither twice a year, at the Feast of the Exaltation of the Cross, and again at the Feast of the Finding of the Cross (which had been grafted on to the old pagan Beltane). The south side would therefore have been an inconvenient site for the monastery. Indeed, the practice of veneration continued long after the Reformation, and as late as 1601 the Minister and Bailies of Peebles report to the Presbytery that at this Beltane "there was no resorting of the people into the Cross Church to commit any sign of superstition there." At the Reformation the monastery was dissolved; and the Cross Church, in succession to that of St Andrew, became the parish church. It was abandoned in 1783 for a new church, built on the Castle Hill at the west end of the High Street. Connected with the monastery was an almshouse and chapel of the Virgin. This almshouse formed a branch establishment of the principal hostel at Eshiels, near Horsburgh Castle—the

Hospital of SS. Leonard and Lawrence, which provided for the pilgrims who journeyed to Peebles from the east.

Dr Gunn, author of the *Books of the Church*, has supplied the following useful summary:

Early Church of St Mungo unrecorded.

St Andrew's 1195. Burned 1549. Abandoned 1560.

Cross Church. Founded 1261. Its Monastery 1473. Dissolved 1560. Parish Church in succession to St Andrew's 1560.

St Mary's. Founded 1363. Used as an Occasional Chapel of the Reformed Faith 1560–1780 (St Mary's stood west of St Andrew's).

Chapel of the Castle of Peebles, *c.* 1153 to 1305.

Chapel and Hospice of SS. Leonard and Lawrence at Eshiels, *c.* 1300–1560.

Lyne Church, still in use, is situated on a gravel moraine east of the Roman Camp. The building measures only 47½ feet by 15 feet, and was built in 1644 by the Hay of Yester who was the first Earl of Tweeddale, on the site of an earlier church.

Stobo Parish Church, a Norman structure, consisting of three parts—tower, nave, chancel—the work of different periods, had considerable alterations made upon it in the sixteenth and seventeenth centuries. The most serious injury inflicted on it was the entire destruction of the Norman chancel arch by the substitution of a modern pointed one when the building was restored in 1868. The sixteenth- and seventeenth-century features consist of a south porch, and a north aisle, which was barrel-vaulted, but is now in ruins. The belfry is of late design, as is also the roof. After the Reformation some of the doors and windows were built up, and the walls plastered. In 1868 an old monumental tomb with canopy was removed, and two Norman windows were discovered.

The Chapel of St Mary's, in Yarrow, situated on a terrace of rock south of Copper Law, about 200 feet

Parish Church, Stobo

(*Drawn by Mr Alex. Blackwood*)

above the level of the loch, has left no traces except a small mound, not over 20 feet square, in the north angle of an enclosure. The oldest name of the church was St Marie of Fairmainshope, and in later times, St Marie of the Lowes, *i.e.* Lochs. According to the ballad *The Douglas Tragedy*, Lord William and Lady

Margaret were buried in the church; and according to the ballad *The Gay Goshawk*, another Lord William in this church roused his lady love from her death-like slumber on her bier. In 1559 the church was attacked by 200 men of the clan Scott, in search of their enemy Sir Peter Cranston, an incident commemorated in Scott's *Lay of the Last Minstrel*.

The site of the primitive church of Selkirk is unknown; the Abbey, however, begun by David I, is supposed to have been at the corner of High Street and Tower Street. A church was built in Selkirk in 1511–12, and another in its place in 1747. It was in the latter church, now a ruin, that the panels of the front gallery were ornamented with pictorial emblems of the various crafts of the burgh, whose deacons and quartermasters occupied the front seats of the gallery. The figure of Justice blind-folded with scales in her hand, and the motto "A false balance is an abomination to the Lord," advertised the piety and the integrity of the Merchant Company. The Tailors represented our first parents making clothes for themselves; the Souters showed a fellow of the order of St Crispin measuring a lady's foot, the explanatory legend being: "How beautiful are thy feet with shoes, O Prince's daughter."

17. Architecture—(*b*) Military: Castles and Peels.

The early castles of Peebles and Selkirk, as in other parts of Great Britain, were at first palisaded earth-works upon which were erected strongholds of timber. Hence Peel, which at first meant a wooden stockade, from the French *pel*, Latin *palus*, a stake, came to designate a fortification with a building inside it, the enclosure as distinct from the building being known as the *barmkyn*. This wooden building was strengthened with an exterior coating of turf and clay. To prevent this wall of turf and clay from collapsing, the rigid structure of timber was built with its four sides sloping inwards, and when stone and lime were substituted for wood and turf the pyramidal form was preserved. In 1535 every landed Borderer possessing £100 worth of land was compelled by law to build a *barmkyn* of stone and lime upon his heritage and lands, with a tower in the same if he thought fit. It was at this time, therefore, that most of the Border keeps of stone and lime were built.

Of the first period (1200–1300) of military architecture in Scotland, no examples exist in Peebles or Selkirk. A distinct break takes place between the thirteenth- and the fourteenth-century type of castle. The country had been impoverished by the Wars of Independence. Besides, Bruce's policy was to build small and inexpensive strongholds, easy to replace and of little value to the English invader. The second period (1300–1400) is, therefore, characterized by small keeps, simple towers; later by keeps of L-shaped plan; and still later, or in the case of wealthy owners, by keeps of E (courtyard) plan. Tinnis Castle, near Drummelzier, so like a robber's castle on the Rhine, built in this century, is exceptional in having four round towers, one at each corner, united by curtain walls. Little remains of it except the foundations.

Neidpath Castle was originally a peel tower, dating probably from the twelfth century. It belonged to the Fraser family and in the fourteenth century came into the hands of the Hays, afterwards earls of Tweeddale. In 1650 it was fortified by John Lord Yester, and besieged by Lambert. The castle, which is of L-shaped plan, is picturesquely situated in a wooded gorge on a rocky prominence overlooking the Tweed winding its way into the valley as it opens out towards Peebles. The walls, which form two oblique angles, are 10 to 11 feet thick. The original door was on the south or precipitous side above the river, and the upper floors were reached by a spiral stair. The tower is divided into two principal compartments by a vault. There is also a vault near the level of the parapet, and probably another carried the roof. Each principal compartment was divided once more into two by wooden floors. The great hall was on the second floor, immediately above the central vault, and was 40 feet long by 21½ feet broad. The corners of the building are all rounded, and the parapet, also rounded, has no projecting bartizans. In the seventeenth century the castle was greatly altered by the second earl of Tweeddale. A courtyard was made to the front, east side, the entrance changed to the centre of this front, a wide staircase introduced, the top storey heightened, the battlements raised so as to contain small apartments, and the parapet fronting the courtyard left open, which was

Neidpath Castle, Peebles

probably the balcony whence the "Maid of Neidpath" viewed the return of her lover, whose failure to recognise her broke her heart.

The third period (1400–1542) still had its simple keeps, of which Newark Castle is a fine example; keeps with one or two wings; and keeps enlarged into castles surrounding a courtyard.

> "Newark's stately tower
> Looks out from Yarrow's birchen bower,"

four and a half miles from Selkirk. In contrast to an older castle, Newark, completed for James III in 1470, means "New Work," and is in a better state of preservation

Newark Tower

than the other strongholds in Yarrow. It was a royal hunting seat in the times of the Stewarts. After the battle of Philiphaugh, 100 prisoners were shot in its courtyard; and it was occupied by Cromwell in 1650. The Duchess of Buccleuch, wife of Monmouth, resided here after his death, and it is during her time and in this castle that Scott makes the "Last Minstrel" sing his *Lay*. The castle is an oblong keep, 65 feet by 40 feet, with walls 10 feet thick and about 84 feet in height. It is surrounded by a barmkyn of irregular shape, about 150 feet square. The first floor is noticeable as it has the hall at one end, and the kitchen at the other with a great fireplace having a seat-cupboard and two mural closets.

To the fourth period (1542–1700) most of the strongholds in Peebles and Selkirk belong. These were mostly abandoned in the seventeenth century or developed into mansion houses. The castles which belong to the period are: Thirlestane, Gamescleuch, Dryhope, Blackhouse, Kirkhope, Oakwood, Barns, Castlehill, Posso, Horsburgh, Nether Horsburgh, Hutcheonfield—all simple keeps; Buckholm, Drummelzier, Cardrona, Haystoun House, have an additional wing added to one end of the main block.

Drochil Castle is an example of the Z plan, having a tower at two of the diagonally opposite angles of the rectangular block so that its defenders might sweep with fire all its four sides at once. The castle has a magnificent situation near the junction of the Tarth and the Lyne. It commands views northwards up the Lyne, westwards up the Tarth, south and east down Lyne valley towards the Tweed valley and the hills behind Hundleshope. The castle is a transition building, and marks the change from the military peel tower with single tenement rooms to a double tenement building in which the military are less pronounced than the domestic features. The towers, for example, are small compared with the size of the building and the shot-holes have been made for musketry, not for cannon. A corridor 12½ feet wide on each storey divides the building into two blocks. The south block now consists of only one storey, but from the northern block it can be seen that the castle had four storeys with attics, each storey being reached by a circular staircase, which began on the ground floor at the front entrance. There were two entrances one at each end of the gallery on the ground floor, the west being the main one. Above this entrance are still to be seen in the tympanum the initials J. D. (James Douglas), the heart and the fetterlock, a D-shaped hobble for a horse, the badge of the Warden of the Marches. The ground floor contains the vaults and cellars, and in the N.E. angle a large kitchen with an immense fireplace and chimney still intact of equal width from floor to roof. The roof of the ground floor is vaulted, and the large hall above this vaulted roof in the south block was the dining room. Although the castle was unfinished when Morton was executed, it seems to have been occupied as a stronghold.

Hallyards is an example of the T plan; Elibank, Whytbank, Torwoodlee of the E or courtyard plan (Scottish type); Traquair of the courtyard plan (Renaissance type), while Fairnilee shows development of a keep into a house and mansion.

The situation of the keeps was chosen mainly for the purpose of giving and receiving fire signals. One fire meant that the enemy was approaching, two that he was coming indeed, and four "all burning together like candles" that he was in great force. The signals passed

Elibank Castle

zigzag from one side of the river to the other up the main valley and its lateral streams till, having been seen all up Teviotdale, Ettrick, Yarrow, and Tweeddale, there gathered by early morning as many as 10,000 men at the rendezvous.

The ground area of the peel-towers often did not exceed 20 feet square. Barns is 28 by 20 feet. The hall on the first floor is only 17½ by 14 feet. It is, therefore, not easy to explain how the owner of a small keep found accommodation for his family and retainers. Originally the first floor of the peel would be reached by a ladder, drawn up when the tower was closed. The ground chambers had always stone vaulted roofs. The bastel houses of Peebles, relics of which were to be seen a few years ago, had both stone vaulted roofs and outside stairs corresponding to the ladder of the peel. The entrance to the vaulted chamber of the peel was by a stout wooden door studded with bolts, and often protected by an iron "yett," the horizontal and vertical bars of which were interlaced to give it additional strength. The "yett" at Barns, probably the oldest in Scotland, is an example of this style of grating. In time the outside approach was dispensed with for a narrow spiral staircase from top to bottom of the tower, generally situated in one of its angles and sometimes in the thickness of the walls. The narrow slots in the walls, deeply splayed on the inside, were meant for arrows; the round holes for fire-arms. The outside of the round holes at Drochil are filleted so as to reduce the chance of shots getting inside, but deeply splayed on the inside so as to increase the angle of fire

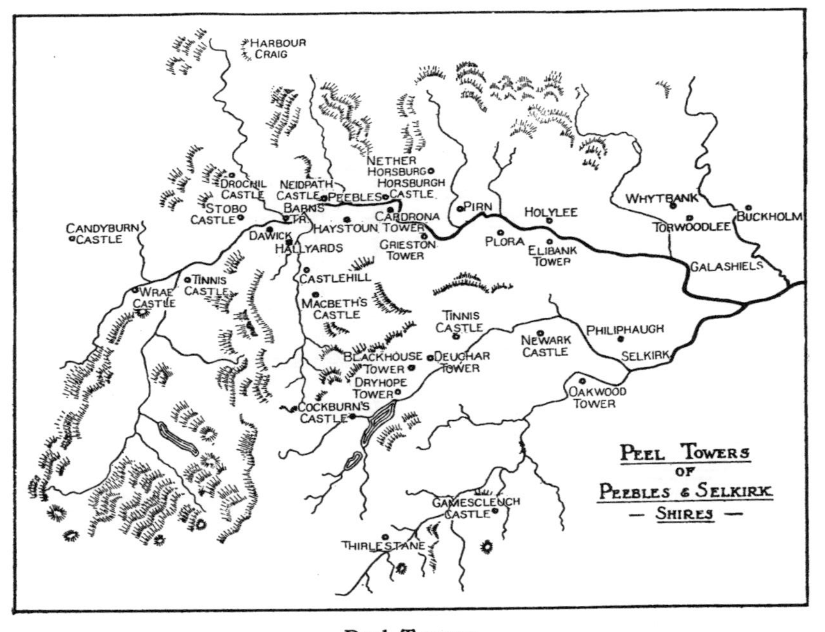
Peel Towers of Peebles & Selkirk — Shires —
Harbour Craig
Candyburn Castle
Drochil Castle
Stobo Castle
Neidpath Castle
Peebles
Barns Tr.
Dawick
Hallyards
Haystoun
Nether Horsburg
Horsburgh Castle
Cardrona Tower
Grieston Tower
Pirn
Holylee
Plora
Elibank Tower
Whytbank
Torwoodlee
Buckholm
Galashiels
Wrae Castle
Tinnis Castle
Castlehill
Macbeth's Castle
Tinnis Castle
Newark Castle
Philiphaugh
Selkirk
Blackhouse Tower
Deuchar Tower
Dryhope Tower
Cockburn's Castle
Oakwood Tower
Gamescleuch Castle
Thirlestane

Peel Towers

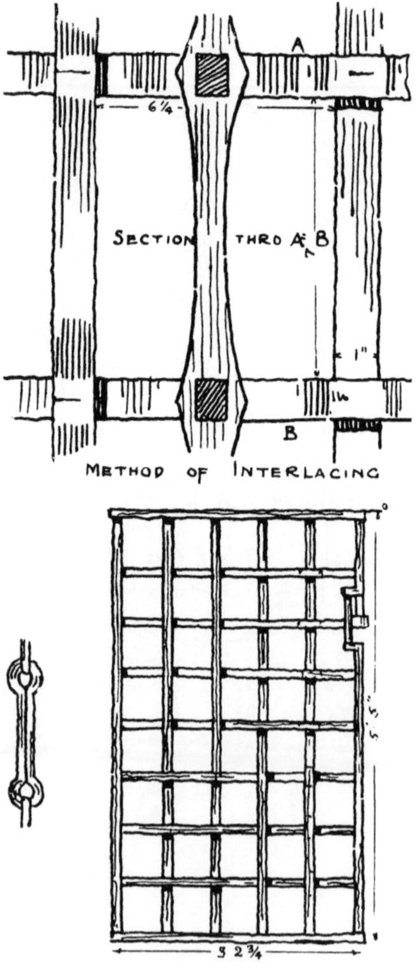

WROUGHT IRON GATE AT BARNS

"Yett" at Barns Tower

for the defenders. The bartizan was the narrow passage between the roof and the battlements. Here the warders kept watch, and here the defence was carried on. Newark and Kirkhope have a bartizan on all sides; Neidpath on west and east. Barns and Oakwood have none. The furnishings depended on the wealth and rank of their owners and on the period. Jamie Telfer had:

> "...naething in his house,
> But ae auld sword
> That hardly now wud fell a mouse";

but the Laird of Torwoodlee was robbed by raiders in 1568 of £1000 in gold and silver, two dozen silver spoons (each two ounce weight), bedding, napery and clothing, abuilzements and plenishing, worth the sum of 5000 merks.

18. Architecture—(*c*) Domestic.

As the need for defence decreased, domestic architecture developed. The transition in Scotland was most pronounced in the reign of James VI. Peels were enlarged into L and E types of building. The castle designed for residence developed, as Drochil; and later the seventeenth-century mansion house, as Traquair and Elibank.

In the Border keep, which had utility stamped upon it, the corbel was designed to bear the parapet; the machiolations to allow guns to be fired from it; the corner turrets to sweep with fire the sides of the building; and the gargoyles to carry off water from the parapets. But as the need for defence disappeared, these useful features of the building were converted to other purposes, or losing their significance, were employed simply as ornament: the turrets became chambers, the corbels were reduced till they became mere chequered bands, as at Drochil, the parapets were absorbed in the walls, and the bartizans disappeared, or became a balcony as at Neidpath. Hence the leading features of seventeenth-century architecture became picturesque turrets cornered out of angles, roofs high pitched with crow-stepped gables, and detail ornamentation with such Norman types as the cable, billet, and dog tooth, as seen at Traquair. The introduction of the Renaissance style was also characterized by a tendency towards uniformity of design, as seen at Fairnilee. Still later, in the eighteenth century, the period began to be marked by the absence of dormer windows, and by the introduction of the unbroken horizontal classic cornice at the eaves, as may be seen at the Whim.

That most interesting mansion, the Glen, was originally a farm-house, to which Playfair, the Edinburgh architect, designed additions. In 1852 Charles, afterwards Sir Charles Tennant, Baronet, of the Glen, purchased the estate and the mansion-house. The house was demolished and the present building, in old Scottish baronial designed by David Bryce, was erected.

The antique aspect of Traquair House or Palace has probably been better preserved than that of any other inhabited house in Scotland. Of Renaissance style and composed of several buildings, it received its present

Fairnilee House

character from John, first Earl of Traquair (1628). The old castle forms the northern portion of the building. The house and offices make three sides of a square, about 100 feet either way, with a beautiful iron railing with

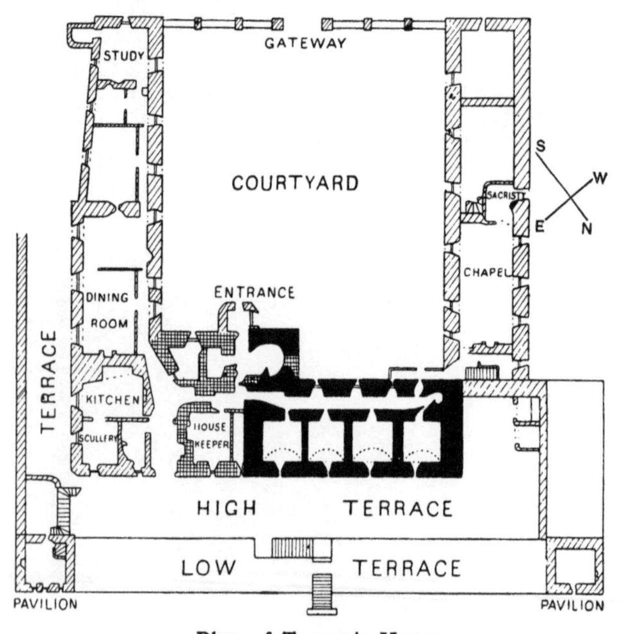

Plan of Traquair House

(*The darkest portions are the oldest*)

stone pillars at intervals and an entrance gateway in the centre. The main building opposite this is four storeys high, with frontage to courtyard and outward or N.E. face, of about 122 feet. The side wings with attics are one storey high. On the N.W. side, which, owing to the fall of the ground has an additional storey, there are the stables and offices, and above, a chapel with sacristy. A high terrace, 17 feet wide, runs along the N.E. side of the building with stairs leading down about eight feet to a lower terrace, at either end of which there is a pavilion with an O. G. roof; a second stair leads down to the banks of the Quair Burn. The building belongs to three periods: first, the old castle on the north; then, the extension (1642) to the S.E., the whole width of the first; finally, the low wings (1695), the terraces and pavilions and the grand entrance gateway. An avenue leads from the front southwards. It is now overgrown with grass, and has been closed for more than 200 years. The famous gateway which opened on to this avenue with its bears rampant and its fine hammered iron railing with ornament of *fleur de lys* is regarded as the prototype of the gateway at Tullyveolan in Scott's *Waverley*. The interior of the house has been little changed since Stewart days. A room on the second floor of the N.E. part of the house has painted decorations on one of its walls—scenes of Eastern life with floral scrolls, and scriptural quotations in old German lettering.

Other buildings, interesting as they are, can only be mentioned. Darnhall with its fine avenue of limes is of Renaissance type, having the appearance of a French château. In the seventeenth century, next to Traquair, it was the finest mansion house in Peeblesshire. Dawyck, surrounded by its beautiful and historic woods and built by Sir James Naesmyth early in the eighteenth century, was in 1864 replaced by the present mansion house of Scottish baronial design. Opposite to it is Stobo Castle, for long the seat of the Montgomery family. Built in 1805–11 by James A. Elliot and situated on an eminence overlooking the Tweed, it presents a bold and striking effect. Halmyre House, Scottish baronial, near the Dead-

Traquair House

burn, was originally a fortalice, part of which is preserved in the lower storey. Lamancha, formerly the Grange, was built by Robert Hamilton in 1663. It was sold to the Dundonald family and its name changed to Lamancha by Alexander Cochrane, son of the eighth Earl of Dundonald, and an Admiral of the Fleet. The Whim,

Renaissance, built by Archibald Earl of Islay (1730), is a massive square, three-storey house. Macbie Hill, adjoining Halmyre, was in the sixteenth century a Border keep. At this period it was known as Coitcoit, according to Nennius the place of King Arthur's seventh battle. The house, whose name was softened to Coldcoat (Coudcoat), was purchased by William Montgomery of Ayrshire

Stobo Castle

and the name changed by him to Macbie Hill, which became the original home of the Montgomeries of Peeblesshire. Spitalhaugh, Scottish baronial, came into the possession of the Fergusson family in 1833, after having passed successively through the hands of the Douglases, the Hays, and the Murrays of Blackbarony.

Returning now to Selkirkshire, we must note Ashiesteel on the south bank of the Tweed, between Walkerburn

Ashiesteel House

and Clovenfords. It was originally a peel, and then a "decent farm house." It is now a low straggling white-washed building, considerably enlarged since Scott occupied it. The older walls are extremely thick. In the grounds is the "Shirra's seat," where Sir Walter Scott wrote much of *Marmion*. Fairnilee, dating back to the fifteenth

Bowhill, Selkirk

century when it was held by the Douglases and the Kerrs, in 1700 came into the hands of Robert Rutherford, one of whose daughters was the famous Alison. The house is a long parallelogram, with entrance door in centre and turrets at each end, ornamented with dog tooth and other "revived" ornaments. Other mansion houses in the vicinity are the new mansion house of Fairnilee, on the opposite side of the Tweed, and the old Castle of Torwoodlee, the scene of one of the last Border raids in 1568. The Haining, near Selkirk, built like an Italian *palazzo*, is one of the finest mansion houses, and is surrounded by perhaps the most beautifully designed gardens and policies in the south of Scotland. The grounds are ornamented with statuary by Canova, and the design of house, gardens, terraces, lake, parks and woods combined with picturesque surroundings forms a most harmonious composition. Philiphaugh in 1792 was an old house with columbarium, orchards and planting. The modern mansion, Scottish baronial, is situated at the foot of a beautifully wooded hill. It has fine terraces along its front, whence extensive views may be had of Yarrow and the country beyond. Bowhill, a name dear to every lover of Scott and the residence of the Dukes of Buccleuch, is built in Renaissance style. Previous to 1455 it belonged to the Douglases, and in the eighteenth century it was acquired by the Dukes of Buccleuch. Duke Charles extended the house and gave it its present appearance. Scott with the affection of a retainer has made the setting of Bowhill for ever famous:

> "When summer smiled on sweet Bowhill,
> And July's eve, with balmy breath,
> Waved the blue-bells on Newark heath;
> When throstles sung on Harehead-shaw,
> And corn was green on Carterhaugh,
> And flourished, broad, Blackandro's oak,
> The aged Harper's soul awoke."

The mansion of Thirlestane, home of the famous Napier family and erected in 1840 in Scottish baronial design, is finely situated amongst lofty plantations on the watershed between Yarrow and Ettrick about two miles above Tushielaw.

19. Communications—Past and Present.

In early times communications between different localities followed the valleys and rivers. In Peebles and Selkirk, then, we find the main roads lying in the longitudinal valleys and the chief transverse valleys. Starting from Galashiels the longitudinal routes stretch up the valleys of the Tweed, Yarrow, and Ettrick to the sources of these streams, and then cross the watershed into Annandale, Eskdale, or Clydesdale. The main road from Galashiels *via* Peebles to Broughton, where the road turns to the left up Tweedsmuir and, crossing the watershed into Annandale by the Devil's Beef Tub, continues through Moffat. Turning east, it follows the Moffat Water to the watershed at Birkhill, descends into Megget, a side valley opening into Yarrow at Capper-cleuch, thence to Tibbie Shiel's inn with Selkirk on the right, and on to Galashiels. Starting once more at Galashiels, the main Carlisle route passes up the valley of Ettrick to Selkirk, and crosses Teviot watershed by Ashkirk to Hawick. A parallel route follows the valley of the Ettrick and, passing up Tima Water, crosses the boundary into Eskdale down to Langholm. Numerous cross roads join these longitudinal routes, over the various watersheds: (1) Tweed, Yarrow and Ettrick; (2) Tweed and Forth; (3) Ettrick, Teviot and Solway; (4) Yarrow and Ettrick.

The obvious route by valley and river was in early days often departed from. The hillsides were chosen, sometimes because drier than the flooded or marshy

Cacra Bank, Ettrick

Route between Ettrick and Teviot (Borthwick Water)

bottoms, sometimes for scouting or for safety, sometimes for other reasons. Let us trace some of these roads. The road over the bridge connecting Pirn Hill with Caerlee Hill passes through the Glenormiston Estate along the south-west slope of Lee Pen to Nether Horsburgh. The road between Peebles and Edinburgh up the lateral valley of the Eddleston water proceeded up hill to Venlaw House. Thence with occasional descents, it passed along the ridge of heights flanking the valley till it crossed the boundary between Peeblesshire and Midlothian, near Portmore. The steepest ascents were at Venlaw and Windylaws; and four horses were required to draw an ordinary travelling vehicle along this road, the rate of progress being three miles an hour. The present road was made in 1770. The old Neidpath road struck up the slope towards Jedderfield, skirting the heights till nearly opposite to Edderston farm, where it came down to the present level. It was probably on account of this difficult road by Neidpath and the want of bridges on the lower part of the Lyne that the old route between Tweeddale and Clydesdale in the seventeenth century came by way of Broughton and Drummelzier. This road crossed the Tweed above Drummelzier by a ford, and was thereafter continued through Manor parish and over the Sware "or Swire" to Peebles. Minchmoor road cuts directly by Traquair over the watershed between Tweed and Yarrow in a line for Selkirk, whereas the present route follows the valley to Caddonfoot and Yair Bridge round behind Sunderland Hall and thence across the Ettrick. The Minchmoor track, which is now a bridle path, has branches leading towards Yarrowford on the right and Ashiesteel on the left, while the main track descends into the valley behind Philiphaugh Farm. The road intersects "Wallace's Trench" and enters Selkirkshire 1800 feet above sea-level. Near the summit behind Traquair it passes a spring called the "Cheese Well," haunted by the fairies. Along the

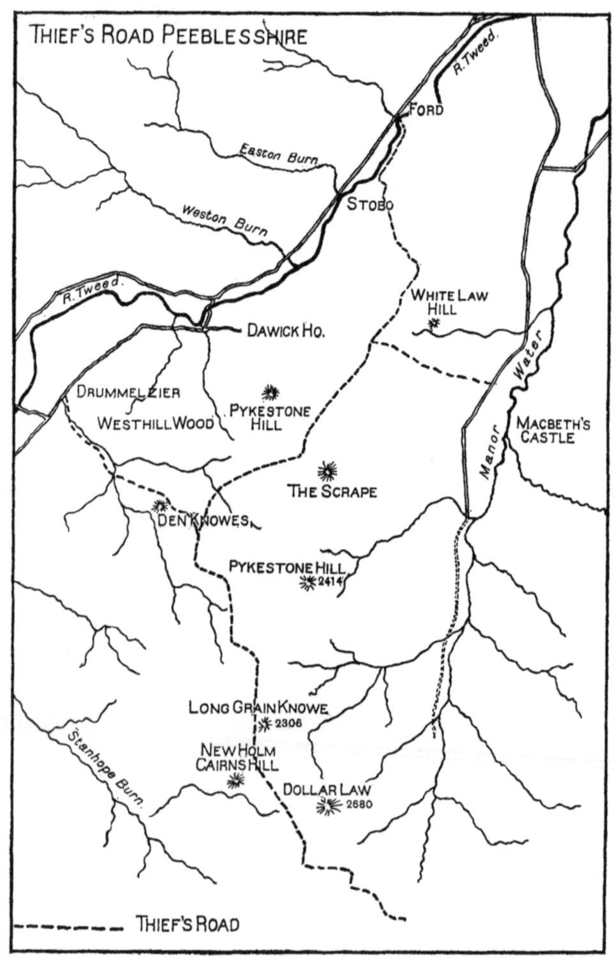

The Thief's Road

Minchmoor road the Peebles millers in the olden days conveyed supplies of meal on pack-horses to Selkirk. In 1769 the Earl of Traquair, on applying to the Peebles Town Council for a subscription to assist in building a bridge over the Quair, astutely reminded the Council of this fact, and was rewarded with the sum of six guineas. "Minchmoor" in Dr John Brown's *Horae Subsecivae* forms the subject of one of his most delightful essays.

There are also transverse hill-roads running mainly north and south over the watersheds. The Drove Road enters the county of Peebles in the north-west corner of Linton parish, near the Cauldstane Slap, crosses Hamilton Hill north-west of the town, passes through Peebles by the "Gipsies' Glen," runs along the ridge between Tweed and Glensax, and descends behind the Glen, continuing thence towards Yarrow and the Borders of England. Such roads in ancient times were exempt from the burdens affecting either parish or turnpike roads, and on passing through Peebles the cattle or sheep with their keepers were permitted for a small fee to rest on what was once known as the Kingsmuir, a spot now occupied by the Caledonian Station. Another well-known road over the backbone of the country, further up the valley, is the Manor Road following the straight valley right up to the steep ridge of Shielhope and Norman Law. Thence up the burn by Bitch Craig (1600 feet), it reaches St Mary's Loch. Other roads of the sort are numerous. But next to Minchmoor the most famous of all these hill roads is the "Thief's Road." This is a broad, flattened, well-marked track without dyke or ditch, so called because it was used by the Border thieves who came and went between the upper reaches of Ettrick and Tweeddale. From the Merecleugh Head or Rodono Hill it passes by the Craigierig Burn, Dollar Law and Scrape to Stobo, a branch leading off to Drummelzier. Below Stobo it crosses the Tweed, and it is said that

Bridge at Ettrick Bridge End

it can be traced through the Pentlands into Midlothian. From Rodono Hill it passes over to Ettrick, where it is known as the "Bridle path," and probably leads into the wilds of Liddesdale. The track is sometimes known as the "King's Road," because James V went by this route to arrest William Cockburn and Adam Scott.

In early days numerous Acts of Parliament were

passed to improve the roads. According to Boston the roads in Ettrick in the middle of the seventeenth century were little better than the channel of a river, being impassable by travellers on horseback, and altogether impracticable to wheeled carriages. In 1719 all the able-bodied men in every district had to give six days' labour in improving the highways. Roads made or improved by this means were called "Statute Labour Roads." But it was not till the close of the eighteenth century that roads and bridges were put into a proper condition. This was done by the Turnpike Act of 1751.

Bridges more than roads appealed to the liberality of individuals and churches in early times, and their erection was sometimes due to pious founders or to the vows of travellers. The first bridge over Ettrick was built at Ettrick Bridge End as the result of a vow by Wat o' Harden. A captive child was drowned as he crossed the ford on his return from a raid, and he vowed to build a bridge so that the one lost life might be the means of saving hundreds. On a stone in this bridge was carved the Harden coat-of-arms: a crescent moon with the motto *Cornua Reparabit Phoebe.* Part of this bridge fell in 1746, and was demolished in 1777 by a flood. A new bridge was built half a mile further up, and the stone with the Harden coat-of-arms transferred to it. Peebles

Old stone with Harden's crest

bridge, built of wood, towards the end of the fifteenth century, was a century later rebuilt of stone. In 1834 it was widened, and in 1890 it was re-built a second time. The various stages of its growth can be seen beneath the arches. At one time Peebles bridge and Berwick bridge were the only two over the Tweed from Peebles to Berwick. One of the largest single-span bridges in Scotland is that over the Tweed at Ashiesteel. Manor bridge at Manorfoot was built in 1702 out of the vacant stipend of the parish, "a most necessar pious use." The inscription states that the bridge was erected by Lord William Douglas, but omits to mention that it was done out of church property. Selkirk bridge over the Ettrick was built in 1778 and enlarged 1881.

Selkirkshire has only one short branch railway line (6¼ miles) joining the Midland route at Galashiels. Peeblesshire has three branch lines, one connecting with the N. B. R. up the Eddleston Valley at Millerhill; the other connecting with the C. R. up the Tweed Valley, *via* Broughton, at Symington; the third connecting Leadburn with Dolphinton on the boundary between Peebles and Lanark. The N. B. branch to Peebles is continued to Galashiels *via* Innerleithen.

20. Administration and Divisions.

Scotland in the twelfth century was divided into twenty-three sheriffdoms, of which Peebles and Selkirk were two. The sheriff, who was generally some high nobleman, was responsible to the King for law and order in his district. The sheriff frequently delegated the active part of his duties to a deputy, and the honorary office as a rule became hereditary. For many years the Murrays of Philiphaugh were hereditary sheriffs of Selkirkshire. When, therefore, in 1747 hereditary jurisdictions were abolished, compensation was paid to the persons holding these rights. Murray of Philiphaugh received £4000; and Lord William, Earl of March, as hereditary sheriff of Tweeddale, £3418 4*s.* 5*d.* At the same time the office of sheriff was vested in the Crown, which was empowered to appoint a sheriff-depute (the sheriff principal), who in turn appointed a sheriff-substitute (the resident county magistrate). The appointment of sheriff-substitute has since been entrusted to the Crown. The depute for Peebleshire is also sheriff of the Lothians; and the depute for Selkirkshire combines in his sheriffdom the neighbouring counties of Roxburgh and Berwick.

Previous to 1889 county affairs were managed by the Commissioners of Supply, the Road Trustees, the Local Authority, the Justices of the Peace, the Police Committee. The Local Government Act of that year transferred the powers and duties of these authorities in whole or part to the County Councils. The Commissioners of Supply, appointed originally in 1667, received their name from the fact that they levied and collected the "cess" or land tax as supply to the Sovereign. Prior to 1889 they had also to appoint the county officials and to maintain a force of police. The Commissioners, who generally speaking comprised the landowners of the district, still meet once a year ; but all the business they transact is to elect a convener, and to concur with the County Council in appointing the Standing Joint Committee for Police.

The Lord-Lieutenant is the military representative of the Crown, and it is his duty to select persons for the Commission of the Peace. In this latter duty he is now assisted by a Local Committee.

Peeblesshire contains the following parishes : Broughton, Glenholm and Kilbucho, Innerleithen, Drummelzier, Eddleston, Kirkurd, Lyne, Manor, Newlands, Peebles, Skirling, Stobo, Traquair, Tweedsmuir, West Linton. The Selkirkshire parishes are : Ashkirk, Caddonfoot, Ettrick, Galashiels, Selkirk, Kirkhope, Yarrow and part of Melrose.

Since 1894 Parish Councils have existed for various local purposes. They administer the poor law, levy poor and school rates, take charge of the registration of births, marriages and deaths, and so on. Primary education is managed by School Boards. With the extension of secondary education it was found that the burgh of the parish was too restricted an area for its administration. County Committees, otherwise known as Secondary Education Committees, were therefore instituted, to co-operate with School Boards in the matter of secondary education ; and they also share the management of the training of teachers.

Peebles and Selkirk are ancient royal burghs, managing their own affairs, under royal charter, by provost, bailies and councillors. Galashiels was erected a burgh of barony in 1599, and became a parliamentary burgh in 1868. In 1869 Innerleithen was made a police burgh. The burgh of Peebles was represented in the Scottish Parliament as early as the reign of David II; the burgh

Seal of the Royal Burgh of Peebles, Dec. 15, 1473
(SIGILVM COMVNI VILLE DE PEBILIS)

of Selkirk was first represented in 1469. Neither county seems to have had a member till the seventeenth century. Various fluctuations, both in burgh and in county representation, took place in the seventeenth century and the eighteenth. In 1831 the proposal to unite the counties of Peebles and Selkirk as one constituency was so strenuously resisted by the Selkirkshire Commissioners of Supply that the proposal was dropped. By the Reform Act of 1832 Peebles and Selkirk were merged with their respective counties. In 1868, however, Selkirk, Hawick and Galashiels were formed into the Hawick Burghs, and known as the "Border Burghs," have since then returned one member, while the counties of Peebles and Selkirk were united in one constituency, returning one member.

21. The Roll of Honour.

The typical Borderer was a fighter and adventurer, and out of his deeds of raid and combat grew the Ballad literature of the Border. Most of the great names of the past are therefore associated either with its warfare or its poetry.

Sir Simon Fraser, the friend and probably the kinsman of Wallace, fought at first on the side of the English. But in 1301 he definitely cast in his lot with the Scottish party, and with Comyn in 1303 won the battle of Roslin. In 1304 on Fraser's own estate at Happrew in Peeblesshire, Wallace and he were defeated by the English. In 1306 he fought with Bruce at Methven, where he saved the king's life. Shortly afterwards, having been captured, he was executed in the same horrible way as Wallace, his handsome appearance and noble bearing compelling the pity and admiration of the spectators.

Bruce's supporters, the Good Sir James and William Douglas, the Knight of Liddesdale, have already been mentioned. After Bannockburn Ettrick Forest came into the possession of the Douglases. James the second Earl of Douglas, was the hero of Chevy Chase, and the dead Douglas that won the field. The fourth Earl died at Verneuil, the sixth was murdered in Edinburgh Castle. The eighth was slain at Stirling, and the ninth defeated in battle at Arkinholm by their implacable foe James II.

Sir Walter Scott of Kirkurd in Peeblesshire was laird of Buccleuch in Ettrick, when he fought against the Black Douglases at Arkinholm; and the Sir Walter Scott, who succeeded his father in 1574, became the first peer of the family, as Lord Scott of Buccleuch. Buccleuch, a typical Borderer and the hero of the ballads, *Jamie Telfer* and *Kinmont Willie*, was the man who when asked by Queen Elizabeth how he dared to break into her castle of Carlisle, replied: "Madam, what is there that a man will not dare to do?" Wat o' Harden, the typical Border Freebooter, is associated with Selkirkshire through his marriage with Mary Scott of Dryhope Tower, "the Flower of Yarrow," as famous for her beauty as Wat was for his courage. He was one of the bold band who recovered Jamie Telfer's kye and broke the gaol to rescue Kinmont Willie. His principal residences in Selkirkshire were Oakwood Tower and Kirkhope Tower. His were the spurs, now in the possession of his descendant, Lord Murray of Elibank, which adorned the dish when the larder was empty; and it was his son Willie Scott who, caught by Gideon Murray at Elibank on a reiving expedition, afterwards married Gideon's daughter, Agnes Murray. The story that Willie Scott got his choice of marrying "Muckle mou'd Meg" or being hanged on the gallows tree, was thought to be disproved when their marriage settlement, a document nine feet long, was discovered. But the story and the settlement are not inconsistent; if the hero reluctantly promised marriage to escape a hanging, the promise may not have been fulfilled till the marriage contract was drawn up.

The "Outlaw Murray" of the ballad belonged to what was till recent times the oldest family in Selkirkshire. The ballad is supposed to refer to John Murray, the eighth laird of Philiphaugh, and the scene is Newark. The Scotts and the Murrays were at feud, and they and other enemies were thought to have prompted the king to make his expedition against the outlaws. The Murrays of Peeblesshire, of the same stock as those of the Forest, come most prominently into notice in the sixteenth century. John Murray, the eighth laird of Blackbarony, knighted in 1592, was known as the first in the district to plant trees and build dry-stone dykes. Hence his name of "John the Dyker." His third son Gideon was father to "Muckle mou'd Meg." Although he could not write his own name, he became Treasurer Depute of Scotland. He had a great liking for architecture and building; and during his tenure of office he had all the royal palaces and castles in Scotland overhauled. Having fallen into disfavour with James, he was sent to prison, where he died of a broken heart. Sir Gideon's son was first Lord Elibank, and a great-great-grandson, the Hon. James Murray, was first Governor-General of Canada in 1763. Besieged in 1781 in Minorca, he was offered by the French general a bribe of £100,000 to surrender but contemptuously refused it, and yielded only when his men were dying of starvation. Another Murray, Alexander Murray of Cringletie, served under Wolfe at Quebec, where he behaved with great gallantry. Murray was as modest as he was brave. When Benjamin West was painting the famous picture of the Death of Wolfe, he requested Murray to pose for one of the figures. But Murray's answer was: "No! No! I was not by, I was leading the left." Murray of Broughton, Prince Charles's Secretary, was the ablest administrator, among the Jacobites of the Forty-five, and the arch-traitor of their cause.

From the sixth Lord Napier of Thirlestane sprang many renowned admirals and generals. William John, eighth Lord Napier, fought at Trafalgar and at Fort Roquette, captured a French privateer, at Almeria cut out a French vessel within half range of 50 guns, was made prisoner at Gibraltar, and after more active service returned home to Ettrick, where he betook himself to farming, historical and antiquarian pursuits. Francis, ninth Lord Napier, after a distinguished diplomatic career, was appointed Governor of Madras in 1866, and on the assassination of Lord Mayo became acting Governor-General of India. He was also a renowned writer and orator.

More noted for craft than for courage, and blighted with the fate of the dynasty whose name they bore, were the Stewarts of Traquair. Sir John Stewart of Traquair was made a peer by Charles I in 1628, and took a leading part in the Covenanting "troubles." He refused to risk his life at Philiphaugh; but, commanding a troop of horse in the Civil War (1648), he was captured. Four years afterwards he was released to find that his son had seized his estates. His remaining years were spent in poverty and disgrace. Dying in 1659, he was buried like a pauper, a shoemaker in pity lending his apron for a pall.

George Pringle of Torwoodlee, a scion of the Pringles of Selkirkshire, a well-known Border family, was appointed sheriff of Selkirk by Richard Cromwell in 1659. On the Restoration he was pardoned but heavily fined. He afforded succour to the Covenanters, assisted the Earl of Argyll to escape to Holland (1681) and, being himself charged with complicity in the Rye-house Plot, fled with Patrick Hume. In Holland Pringle was one of the council of twelve for the recovery of the rights and liberties of Scotland, and one of the committee of seven who planned Argyll's invasion. In 1689 he, along with Scott of Harden, represented Selkirkshire in the Scottish Convention which offered the crown to William and Mary. His estates were restored; but, worn out with his hardships, he died the same year.

With a taste for natural science, Mungo Park (1771–1806), son of a Foulshiels farmer, inherited the Borderer's love of adventure. Educated at Selkirk Grammar School, he studied medicine, and sailed as surgeon to Sumatra. In 1795 he went to explore the Niger region.

This made him famous, and his *Travels in the Interior of Africa* is still a classic. He settled in Peebles as a

Mungo Park

medical practitioner, but tired of the life and returned to the Niger, where he was drowned.

Though Michael Scott the Wizard (1175–1235) has only a supposed connexion with Selkirkshire, his name and fame are wedded with its history and literature. As a student of science and magic he had a European reputation:

"When, in Salamanca's cave,
Him listed his magic wand to wave
The bells would ring in Notre Dame."

He was tutor to the Emperor Frederick II, and court physician and astrologer at Palermo. Returning to Scotland in 1230, he died about five years afterwards, and is buried, says tradition, in Melrose Abbey. His reputed abode was Oakwood Tower; but this Border keep was not built till 300 years after the wizard's death. Dante has figured him in Purgatory with his head turned round looking backward because in life he had been a diviner.

The writers of the numerous old ballads are all unnamed save Nicol Burne, author of *Leader Haughs and Yarrow*. He is supposed to have been the foundling whom Mary Scott discovered forgotten amongst the baggage after the return of her husband, Wat o' Harden, from a raid in Northumberland.

"He nameless as the race from which he sprung
Saved other names and left his own unsung."

But, known or unknown, the succession of poets has never failed. Robert Crawford (1695–1732) was author of *Tweedside* and of *The Bush aboon Traquair*. Hamilton of Bangour (1704–1754) wrote the *Braes o' Yarrow*, the measure of which was imitated by Wordsworth in the

Yarrow poems. Willie Laidlaw (1780–1845), born at Blackhouse in Selkirkshire, was the author of the pathetic lyric *Lucy's Flittin.* Alison Rutherford (Mrs Cockburn), born at Fairnilee in 1712 and educated in Edinburgh, where she soon became renowned for her wit and beauty, was in very truth a nymph of the "Forest" and a "Maid of Athens." She was the authoress of the immortal *Flowers of the Forest.*

Sir Walter Scott, greatest of the Border minstrels and best of men, was, though born in Edinburgh, closely associated with Selkirkshire, by descent on the father's side from Wat o' Harden and on the mother's from the Rev. John Rutherford of Yarrow, by his official position as sheriff of the county, and by residence at Ashiesteel. The scenery, the people, the life, the history, the traditions of Selkirk and Peebles—all influenced Scott and Scott's work. "If no country ever owed so much for its fame to one man as Scotland to Sir Walter Scott, no part of it has so earned distinction through his notice as Selkirkshire." *The Minstrelsy* is full of Selkirk influences, *Marmion* was written and *Waverley* was begun at Ashiesteel. Scott's pictorial power is finely displayed in local scenes as: "Tweedside in November" (*Marmion*, introduction to canto i); "Yarrow" (introduction to cantos iv and v); "A Snowstorm amongst the Hills" (canto iv); and, one of the best, "St Mary's Loch in Calm" (introduction to canto ii). His novels are full of allusions to places and persons in the shires, and two of them, *St Ronan's Well* and *The Black Dwarf*, deal especially with the district.

James Hogg, "the Ettrick Shepherd," was born in 1770, and, with few and short migrations, lived in

Hogg's Monument at St Mary's Loch

Selkirkshire all his life of sixty-five years as a shepherd and as a sheep farmer. He said he preferred a

Border fair to a King's coronation. His first important work was *The Mountain Bard*. His masterpiece is *The Queen's Wake*, but his exquisite song, *When the kye comes hame* must not be forgotten. Though far below Burns as a poet, "there is a marked individuality in the shepherd's songs and poems; he was a singer by genuine impulse, and there was an open-air freshness in his note." James Nicol (1769–1819), minister of Traquair, wrote *Where Quair rins sweet amang the Flouirs*; and Thomas Smibert (1810–1834), a doctor and a native of Peebles, the *Scottish Widow's Lament*. Professor John Wilson, "Christopher North" (1785–1854), was author of 39 out of 70 of the *Noctes*, and the friend of Wordsworth and of Hogg. Thomas Tod Stoddart (1810–1880) in his fishing songs praises Lyne, Manor, Yarrow, Gala, Tweed. Thomas Pringle (1789–1834) wrote his *Autumnal Excursion*, inspired by a visit to St Mary's Loch. Another and finer *Bush aboon Traquair* was written by Principal Shairp (1819–1885). The Rev. Dr Russell in his *Reminiscences* and the Rev. Dr Borland in his Anthologies have carried on the literary tradition of Yarrow. John Veitch, a disciple of Scott and Wordsworth, was born in Peebles, 1829, and died there, 1894. He was professor of Logic at St Andrews and at Glasgow. Among his writings associated with his native district are *Tweed and other Poems* and his *History and Poetry of the Scottish Border*—the standard book on the subject. James Brown (1852–1904), a Selkirk manufacturer, under the *nom de plume* of J. B. Selkirk, wrote *Selkirk after Flodden* and *O Yarrow garlanded with rhyme*. As a poet and a

9—2

man of letters, Andrew Lang (1844–1912), is the most distinguished son of Selkirkshire in modern times. Born at Selkirk, where he spent his childhood, he early displayed a bent towards literary pursuits. In range and productiveness he has had no rivals in Great Britain, and has even been seriously regarded as a society of authors. History, poetry, biography, belles-lettres, and comparative religion he treated with learning, liveliness and interest. His love for his native district has been beautifully expressed in *Twilight on Tweed* and *Sunset on Yarrow.*

The brothers William and Robert Chambers, the publishers, are the most eminent men of letters of modern times belonging to Peeblesshire. They were born at Peebles, William in 1800, Robert in 1802. In 1832 William started *Chambers's Edinburgh Journal.* In 1859 he founded the Chambers Institute in Peebles. He was Lord Provost of Edinburgh, and carried out at his own cost the restoration of St Giles' Cathedral. He died in 1883. His *History of Peeblesshire* is the standard book on the subject. Robert's *Vestiges of Creation* was an anticipation of Darwin's *Origin of Species.* His numerous other volumes include *History of the Rebellions in Scotland*, *Dictionary of Eminent Scotsmen*, *The Life and Works of Robert Burns.* Henry Calderwood (1830–1899), minister in Edinburgh and professor of Moral Philosophy, wrote *Philosophy of the Infinite* and *Mind and Brain.* James Nicol, professor of Natural Philosophy in the University of Aberdeen, was the first discoverer of graptolites in the greywacke of the district. James

Andrew Lang

Wilson, editor of the *Border Advertiser*, contributed to the elucidation of Professor Lapworth's theory regarding the Silurian formation of the Southern Uplands.

In law and politics there are also eminent names. Andrew Pringle (Lord Alemoor), the son of John Pringle of the Haining, a senator of the College of Justice, was successively Sheriff of Selkirk, Solicitor-General for Scotland and a judge of the Court of Session. Sir James Montgomery's name is honourably associated with land reform in the eighteenth century. The second son of William Montgomery of Macbie Hill, he was successively Sheriff of Peeblesshire, Solicitor-General, Lord Advocate, and Baron of the Exchequer of Scotland. In 1745 he purchased the estate of Stanhope and became an "Improver." Later he bought the Whim from the Duke of Argyll, and found as much wine in the cellar as paid for the estate. He was the author of the Entail Act, so advantageous to agricultural progress in Scotland. Montgomery also took an active part in the Parliamentary abolition of serfdom in Scotland. Macqueen of Braxfield was of a different type. A ferocious partisan in politics, he acted as a sort of Judge Jeffreys for the reactionary government of the period, *circa* 1793, in its efforts to repress the movement for political reform. Forbes Mackenzie of Portmore was M.P. for the county in 1830, and was responsible for the Forbes Mackenzie Act, the first important measure of licensing reform, which would have been unnecessary had every Scottish hostess followed the precepts and practice of "Meg Dods" of the Cleikum Inn at Peebles, who—according

Professor George Lawson

to Scott in *St Ronan's Well*—discouraged late hours and deep potations.

Some of the most distinguished names in the history of the Scottish Church for the past 400 years are associated with the two shires. John Welsh (1568–1622), the famous preacher, in his youth consorted with the thieves of Liddesdale. He was minister of Selkirk, of Kirkcudbright and of Ayr. After imprisonment, he was banished and went to France, where he became minister to the Huguenots at St Jean d'Angley. Another famous preacher was Thomas Boston, appointed to Ettrick in 1707. Notable books of his are the *Fourfold State*, *The Crook in the Lot*, and his autobiography. Professor Lawson was born at Boghouse, Peeblesshire, and for fifty years had charge of the Secession Church at Selkirk. One of his students, John Lee, joining the Church of Scotland, was appointed minister of Peebles, then professor of Church History at St Andrews, and finally Principal of Edinburgh University. Professor John Ker (1816–1886), born at the Bield, Tweedsmuir, became one of the most brilliant preachers of the United Presbyterian Church.

Some of those who did much to promote the woollen industry in the district were Dickson of Peebles, the first manufacturer to make shepherd-tartan trousers, the origin of checked Tweeds; Murray of Galashiels, who brought Australian wool into vogue; Mercer, "the enterprising pioneer of the local industry" in the use of machinery; and George Roberts, who introduced a set of carding engines, an American invention.

22. THE CHIEF TOWNS AND VILLAGES

(The figures in brackets after each name give the population in 1911, and those at the end of each section are references to pages in the text.)

A.—PEEBLESSHIRE.

Broughton (pa. 668), situated where the Tweed flowing north from Tweedsmuir turns eastward. There are many British forts in the neighbourhood and relics of the bronze period have been frequently found. (pp. 8, 10, 13, 58, 76, 79, 111, 113, 118, 125.)

Cardrona, a small hamlet in the parish of Traquair midway between Peebles and Innerleithen. (pp. 33, 41, 49, 79, 84, 95.)

Carlops, in West Linton parish three miles N.E. of West Linton. Its old name was Carlynlippis and it was from 1334 to 1357 one of the landmarks of the northern boundary of England, which at that time included part of Peeblesshire. "Habbie's Howe" near Carlops in the valley of the Esk is the scene of Allan Ramsay's *Gentle Shepherd*. (pp. 23, 66, 68.)

Drummelzier (pa. 164), three miles S.E. of Broughton Station, has a pre-Reformation parish church. A thornbush near the churchyard marks the traditional burial-place of Merlin the Wizard. (pp. 14, 15, 17, 32, 33, 58, 85, 92, 95, 113, 116.)

Eddleston (pa. 589) is a village 4½ miles N. of Peebles. In the neighbourhood is the beautiful cascade of Cowie's Lynn.

West of the village stands Darnhall, the seat of the Murrays—called formerly Halton, and afterwards Blackbarony. (pp. 11, 58.)

Innerleithen (2547), near the mouth of the Leithen Water, has large woollen mills. Long famous as a summer resort, it had, in the early part of last century, some renown as a watering-place. In early times the church of Innerleithen was dedicated to St Kentigern. Malcolm II bestowed upon it the right of sanctuary because the dead body of his son, who had been accidentally drowned in Tweed, had lain there one night before burial. The Carnegie Free Library is a building of Elizabethan design. (pp. 8, 11, 17, 32, 49, 61, 67, 118, 121.)

Kirkurd (pa. 253), a village about nine miles N.W. of Peebles. (pp. 58, 66.)

Lyne (pa. 125), a hamlet on the left bank of Lyne Water beneath the southern slope of the plateau on which Lyne camp is situated. The neighbourhood is noted for its British forts, pre-historic remains, and the church built in 1644. The pulpit, presented by Lady Yester, is a highly finished piece of woodwork from Holland. (pp. 7, 11, 33, 58, 66, 82, 83, 84, 88.)

Lamancha, a small village in the parish of Newlands, used to be the seat of the Earls of Dundonald. (pp. 32, 106.)

Manor (pa. 261) is a scattered hamlet in the valley of Manor Water. In the churchyard is the grave of the Black Dwarf, who lived in a cottage erected by himself near Woodhouse farm. Posso near the south end of the valley is famous for its falcons. Hill forts are numerous. One, Macbeth's Castle, occupies a *roche moutonnée* in the middle of the valley. South of Posso Craig stood the parish church, known as St Gordian's Kirk, till about the year 1650. In 1874 a cross was erected by Sir James Naesmyth of Posso to mark the spot. Between St Gordian's Cross and Manorhead, a monumental cairn has been erected to the memory of Professor John Veitch "in his favourite valley." (pp. 7, 11, 34, 38, 39, 40, 41, 49, 58, 84, 113, 118.)

Newlands (pa. 590), a hamlet between West Linton and Eddleston. (pp. 38, 58, 120.)

Peebles (5554), the county town and an ancient royal burgh, was in existence before 1195. The old town lay north of the Tweed and west of Eddleston Water. The only part now remaining stretches from Bigglesknowe to St Andrew's Tower.

Queensberry Lodging as possessed by 'Old Q' in the Eighteenth Century

Purchased by William Chambers 1857

The old town was more than once burned by the English, and in the sixteenth century a new town sprang up along the high ridge extending from the site of the parish church to the East-gate. The new town was surrounded by a wall, a portion of which may still be seen on Venlaw Road. Peebles was a famous ecclesiastical centre till the Reformation, and a favourite residence of the Scottish Kings. David II granted it a charter in 1337, and probably David I had done the same. Bruce, having recovered it from the English, demolished its Castle. After the Reformation a number of the nobility and gentry took up their residence in Peebles; but after the Union they left it for London. By the middle of the eighteenth century the spirit of commercial enterprise had awakened in the place, and since the introduction of the Tweed manufacture the town has steadily developed.

In the year 1624 the building, afterwards known as the Queensberry Lodging, was presented by James VI to Lord Yester, ancestor of the Marquis of Tweeddale; and in 1687 became the property of the Duke of Queensberry. There is a current tradition that "Old Q" was born in the Queensberry Lodging. In 1857 it was acquired by William Chambers, who presented it to his native town. By Chambers the building was entirely reconstructed with the exception of the vaulted ground-floor. Recently, through the munificence of Mr Andrew Carnegie, the Chambers Institution was reconstructed and extended. The new buildings, opened in 1912, comprise the Council Chambers, the Town Hall, the Library, the Museum, doubled in size, an Art Gallery and other accommodation. Peebles has also County Buildings, a fine specimen of Tudor design; a High School for Burgh and County; a Hydro; and numerous churches. The old Cross, in High Street, has had an eventful history. There is a golf course, with a fine southern exposure, overlooking town and valley.

The Burgh arms (see page 121) are three salmon naiant counter naiant, with the legend, *Contra nando incrementum.* It was a common jest in more convivial days, when the saying "Peebles for Pleesure" had its origin, to make them "three tumblers." (pp. 1, 2, 14, 15, 17, 30, 33, 35, 61, 64, 67, 69, 74, 75, 76, 77, 85, 87, 98, 111, 112, 113, 115, 118, 120, 121, 122, 127, 131, 132, 134, 136.)

Romanno Bridge, a hamlet in the parish of Newlands,

has famous terraces; whether made by the Britons, or the Romans, or the monks of Newbattle Abbey, is uncertain. Similar terraces occur on Roger's Crag, east of Halmyre. (pp. 11, 13.)

Stobo (pa. 350), a hamlet seven miles west of Peebles. The church, a Plebania or mother church in early times, is mentioned in the Inquisition of David I as having belonged to Kentigern, and in the Peebles Burgh Records as "Saint Mungoy's Kirk of Stobo." John Reid of Stobo, churchman and notary, is one of the poets whom William Dunbar mourns for in his *Lament for the Makaris*:

> "And he [Death] has now tane, last of aw,
> Gud gentill Stobo et Quintyne Schaw,
> Of quham all wichtis hes pete:
> Timor Mortis conturbat me."

(pp. 13, 17, 30, 40, 41, 44, 45, 59, 67, 85, 88, 106, 116.)

Traquair (pa. 559), a well-known hamlet near the Quair Burn opposite to Innerleithen, is famous for its associations with Traquair House and for the song *The Bush aboon Traquair*. (pp. 1, 8, 17, 18, 27, 33, 49, 58, 68, 76, 77, 85, 96, 101, 102, 104, 105, 113, 126, 131.)

Tweedsmuir (pa. 198) is a small hamlet in the upper reaches of the Tweed. The churchyard contains the grave of John Hunter, a martyr for the Covenant. In the neighbourhood is Oliver Castle, built about 1200, the home of the Frasers. (pp. 14, 15, 41, 58, 77, 84, 85, 136.)

Walkerburn (1331), a village in the parish of Innerleithen, founded in 1855 by Henry Ballantyne, in whose memory the Ballantyne Memorial Institute was erected, 1903. On the face of Purvis Hill near Walkerburn is a range of terraces similar to those at Romanno. Opposite to Walkerburn is the Plora glen, in which Hogg's "Bonnie Kilmeny" was spirited away by the fairies. (pp. 11, 32, 61, 107.)

West Linton (pa. 1000) is a favourite summer resort for Edinburgh people, owing to its nearness (sixteen miles) to the capital and its healthy situation, 600 feet above sea-level. The parish church has fine wood-carving. Linton was the first known settlement of the Comyn family in Scotland. West Linton was formerly a burgh of regality, with a baron-bailie and a council of feuars, called the "Linton Lairds." One of these, Laird Gifford, was a noted local sculptor. The finial of the Jubilee clock, representing his wife, is his work. West Linton had become famous for its stone carvers from the time when the builders at Drochil Castle introduced their art to the village. (pp. 13, 23, 34, 35, 40, 49, 58, 66, 67.)

B.—SELKIRKSHIRE.

Ashkirk (pa. 329), a village 5½ miles south of Selkirk. (pp. 8, 22, 58, 111, 120.)

Caddonfoot (pa. 709), a hamlet four miles south-west of Galashiels, is the scene of the old ballad *Katharine Janfarie*, which suggested *Lochinvar* to Scott. (pp. 9, 18, 49, 58, 66, 113, 120.)

Chapelhope is a small hamlet at the head of the Loch o' the Lowes. Near at hand is the statue to James Hogg, "the Ettrick Shepherd." Chapelhope was originally the site of Rodono Chapel, and there are traces of a mote on which the Bailies of Rodono dispensed justice on behalf of the Abbots of Melrose.

Clovenfords, a small village in Caddonfoot parish, has memories of Scott, De Quincey, Leyden, and Wordsworth. In 1867 relics were discovered on Meigle Hill of an old military encampment, comprising scrap iron, broken blacksmith's tongs, and fragments of sheet bronze. (pp. 18, 59, 109.)

Ettrick (pa. 344) consists of a church, a school, a manse and a churchyard. In the churchyard lie buried Boston; Hogg;

Hogg's grandfather; "Will o' the Phaup," a noted athlete; Tibbie Shiel; and Baron Napier of Ettrick. Hogg's birthplace, Ettrick-hall farm, is near the church. (pp. 19, 48, 58, 116, 117, 125, 136.)

Galashiels (14,531), a parliamentary burgh, occupies 2½ miles of the narrow valley of the Gala before its junction with the Tweed. In 1559 it was made a burgh of barony, having then only 400 inhabitants. Towards the end of the eighteenth century Galashiels according to Dorothy Wordsworth was a large irregularly built village, just beginning to assume a "townish bustle." It was at this time, during the Napoleonic Wars, that Galashiels got and took its opportunity to develop its trade. It is now the chief seat in Scotland of the Tweed manufacture. The rapid rise of the trade is marked by the fact that the annual value of its woollen manufactures rose from £1000 in 1790 to £1,250,000 in 1890, when the population was 17,367. The trade depression that followed reduced the population in 1901 to 13,615, after which a revival began. The increased prosperity of the town has been shown not only by the growth of the population but also by the introduction of a costly drainage scheme, the erection of a Technical College, a handsome building of red sandstone in classic style, the opening of a new Secondary School, and the laying out of a new town square. The square includes a fountain with a shaft, surmounting the capital and frieze of which is a reproduction of the town's coat-of-arms—a fox in the attempt to reach some pendent plums, with the legend, "Soor Plums." This is associated with a song the tune of which alone remains. The song, *Sour Plums in Galashiels*, commemorated a defeat inflicted by the natives on the English, who were regaling themselves with the wild plums which grew near the village. Another song connected with the town is *Gala Water*, on which Burns built his beautiful lyric with the refrain "Braw, Braw Lads." These words are appropriately inscribed on the base of the bronze bust of Burns at the foot of Lawyers Brae.

Besides Tweed manufactures, Galashiels has dyeworks, iron foundries, engineering works, and boot factories. (pp. 5, 7, 9, 12, 15, 27, 30, 44, 45, 52, 58, 59, 61, 62, 64, 66, 76, 77, 79, 111, 121, 122.)

Kirkhope (pa. 384) consists of a manse, a farm steading, and the old tower of Kirkhope. (pp. 58, 95, 101, 128, 136.)

Flodden Flag

Selkirk (5886) is a royal burgh and the county town. Notable features are the Old Town Hall with its clock and spire, the statues of Sir Walter Scott and Mungo Park, the New Town Hall, the "Mercat" Cross, the Flodden Memorial. Selkirk Abbey, unfinished, and Selkirk Castle, the frequent abode of Scottish kings, date back to the twelfth and thirteenth centuries. In 1418 the town was burned by the English. At Flodden

Selkirk lost a large proportion of its burghers. Only one of the Selkirk contingent returned. He is figured in the town's memorial of the battle, holding aloft in his hand an English pennon which the Selkirk men won from their foes (see page 73). A flag still preserved is, according to tradition, this very trophy. In 1640 Provost Muthag was slain while defending the burghlands from the aggressions of Ker of Brigheuch, a neighbouring laird. The town's war-song is

> "Up wi' the Souters o' Selkirk
> And down wi' the Earl o' Hume,"

the tune of which is peculiar in ending on the dominant seventh.

The coat-of-arms of the Burgh is a female figure holding a child in her arms, supposed to be the Virgin and the Child, and most likely adopted from the seal of St Mary's Church of Selkirk. The shield with the lion rampant was added probably in the time of James V. The motto is: *Et spreta incolumem vita defendere famam.* (pp. 2, 5, 42, 52, 58, 60, 64, 66, 73, 76, 77, 85, 90, 118, 120, 121, 122, 126, 131, 132, 136.)

Yarrow (pa. 510) consists of a church, a school, a police-station, and a group of houses by the roadside. (pp. 7, 8, 9, 20, 33, 36, 49, 58, 67, 68, 71, 84, 89, 113, 115, 129, 131.)

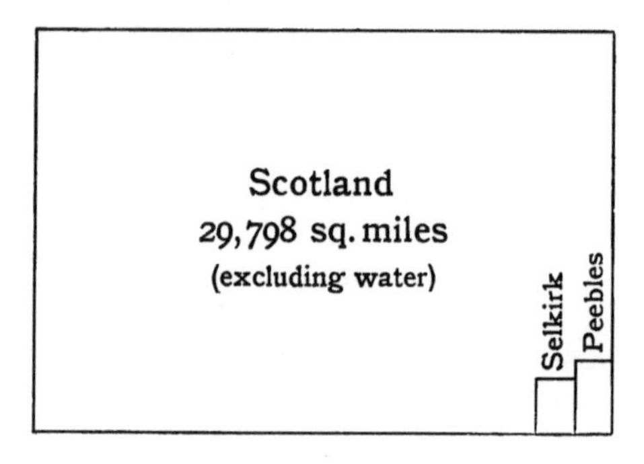

Fig. 1. Areas (excluding water) of Peebles (347 sq. miles) and Selkirk (267 sq. miles) compared with that of Scotland

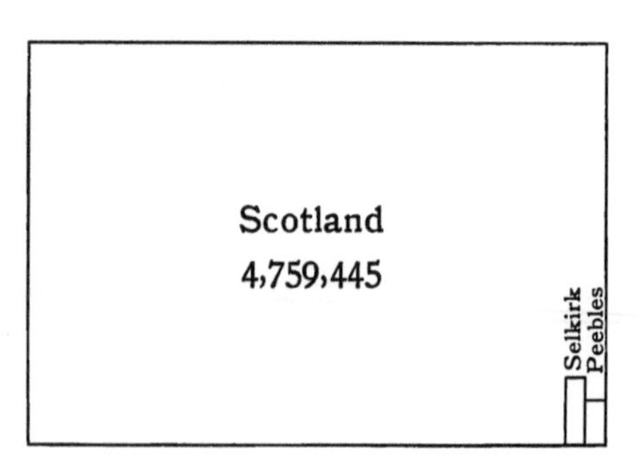

Fig. 2. Population of Peebles (15,258) and Selkirk (24,600) compared with that of Scotland in 1911

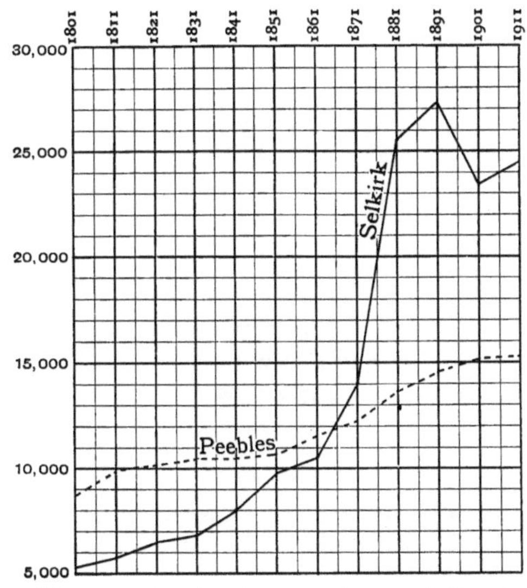

Fig. 3. Diagram showing increase in population in Peebles and Selkirk since 1801

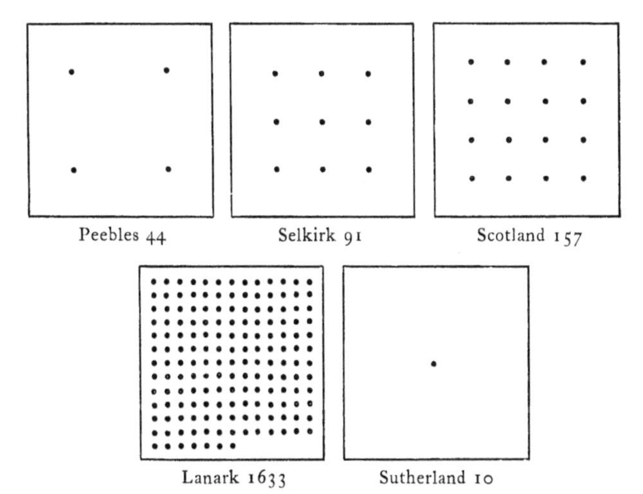

Fig. 4. Comparative Density of Population to the square mile in 1911

(*Each dot represents* 10 *persons*)

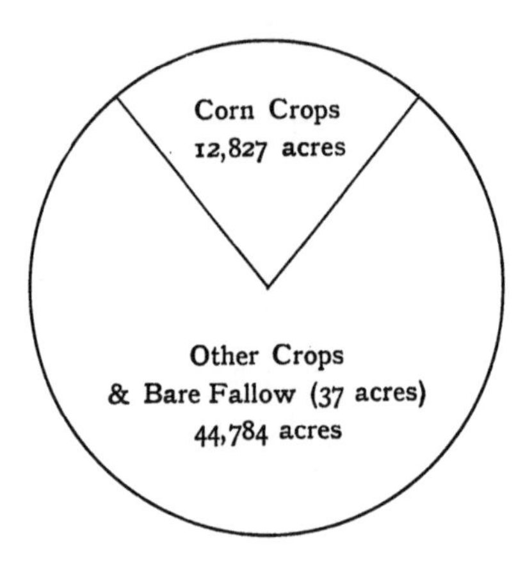

Fig. 5. Proportionate area under Corn Crops compared with that of other cultivated land in Peebles and Selkirk in 1912

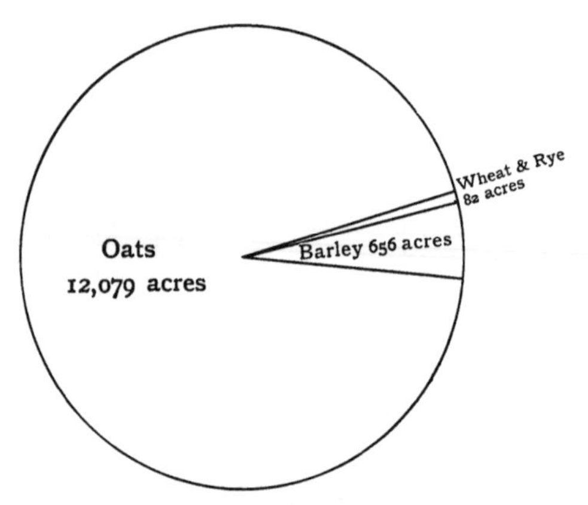

Fig. 6. Proportionate areas of chief cereals in Peebles and Selkirk in 1912

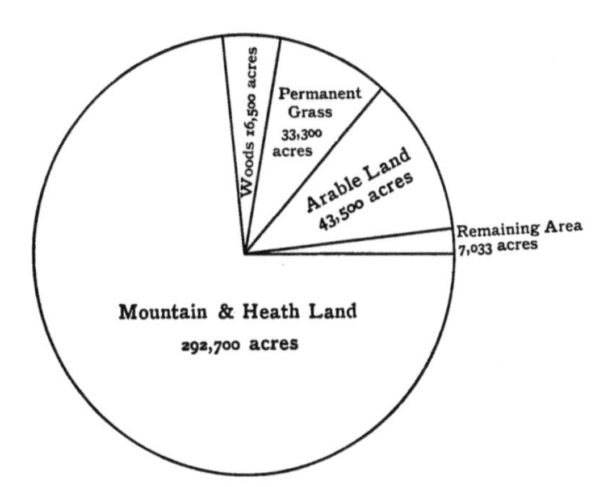

Fig. 7. Proportionate areas of land in Peebles and Selkirk in 1912

Fig. 8. Proportionate numbers of Live Stock in Peebles and Selkirk in 1912

www.ingramcontent.com/pod-product-compliance
Ingram Content Group UK Ltd.
Pitfield, Milton Keynes, MK11 3LW, UK
UKHW042145280225
455719UK00001B/115